Sustaining Moore's Law

Uncertainty Leading to a Certainty of IoT Revolution

Synthesis Lectures on Emerging Engineering Technologies

Editor
Kris Iniewski, *Redlen Technologies Inc.*

Sustaining Moore's Law: Uncertainty Leading to a Certainty of IoT Revolution
Apek Mulay
2015

Sustaining Moore's Law: Uncertainty Leading to a Certainty of IoT Revolution

Apek Mulay

ISBN: 978-3-031-00897-9   paperback
ISBN: 978-3-031-02025-4   ebook

DOI 10.1007/978-3-031-02025-4

A Publication in the Springer series
SYNTHESIS LECTURES ON EMERGING ENGINEERING TECHNOLOGIES

Lecture #1
Series ISSN
ISSN pending.

Sustaining Moore's Law

Uncertainty Leading to a Certainty of IoT Revolution

Apek Mulay

Mulay's Consultancy Services

SYNTHESIS LECTURES ON EMERGING ENGINEERING TECHNOLOGIES #1

ABSTRACT

In 1965, Intel co-founder Gordon Moore, in "Cramming more components onto Integrated Circuits" in *Electronics Magazine* (April 19, 1965), made the observation that, in the history of computing hardware, the number of transistors on integrated circuits doubles approximately every two years. Since its inception in 1965 until recent times, this law has been used in the semiconductor industry to guide investments for long-term planning as well as to set targets for research and development. These investments have helped in a productive utilization of wealth, which created more employment opportunities for semiconductor industry professionals. In this way, the development of Moore's Law has helped sustain the progress of today's knowledge-based economy.

While Moore's Law has, on one hand, helped drive investments toward technological and economic growth, thereby benefiting the consumers with more powerful electronic gadgets, Moore's Law has indirectly also helped to fuel other innovations in the global economy. However, the Law of diminishing returns is now questioning the sustainability of further evolution of Moore's Law and its ability to sustain the progress of today's knowledge based economy. The lack of liquidity in the global economy is truly bringing the entire industry to a standstill and the dark clouds of an economic depression are hovering over the global economy.

What factors have been ignored by the global semiconductor industry leading to a demise of Moore's Law? Do the existing business models prevalent in the semiconductor industry pose any problems? Have supply chains made that progress unsustainable? In today's globalized world, have businesses been able to sustain national interests while driving the progress of Moore's Law? Could the semiconductor industry help the entire global economy move toward a radiance of the new crimson dawn, beyond the veil of the darkest night by sustaining the progress of Moore's Law?

The entire semiconductor industry is now clamoring for a fresh approach to overcome existing barriers to the progress of Moore's Law, and this book delivers just that. Moore's Law can easily continue for the foreseeable future if the chip manufacturing industry becomes sustainable by having a balanced economy. The sustainable progress of Moore's Law advocates the "heresy" of transforming the current economic orthodoxy of monopoly capitalism into free-market capitalism. The next big thing that everybody is looking forward to after mobile revolution is the "Internet of Things" (IoT) revolution. While some analysts forecast that the IoT market would achieve 5.4 billion connections worldwide by 2020, the poor consumer purchasing power in global economy makes this forecast truly questionable. *Sustaining Moore's Law* presents a blueprint for sustaining the progress of Moore's Law to bring about IoT Revolution in the global economy.

KEYWORDS

global economic crisis, counterfeit electronic components, capitalism and the free market, Moore's Law, mass capitalism, semiconductor industry market trend, Make in India, 450 mm silicon wafers, Internet of Things (IoT)

Contents

Foreword

What is commonly known as Moore's Law is a mystery to most of us, but to those who are engaged in the production and marketing of computers and related products the law is an awe-inspiring discovery made by Gordon Moore as far back as 1965. The law is actually a forecast that processor speeds or a computer's processing power will roughly double every two years. Incredible as it is, the forecast has been fairly accurate over the years.

Moore's Law relates not only to technology but also to the economics of the electronics industry. Moore's forecast suggested that computer costs would fall sharply over the coming decades and that computers would be increasingly used in science as well as R&D activity. Now, Mr. Apek Mulay, an electronics engineer, extends Moore's Law to the workings of macroeconomics. He takes a big leap from the economics of the electronics industry to the economics of a nation.

In a relatively short period, Mr. Mulay has become a prolific writer. His first book, *Mass Capitalism: A Blueprint for Economic Revival*, dealt with the relationship between Moore's Law and macroeconomic policy. Now he has written another volume, *Sustaining Moore's Law: Uncertainty Leading to a Certainty of IoT Revolution*, that extends his earlier analysis to many new areas. It is well known that there is hardly any area of technology left untouched by the law. But it is no longer clear if the law's bold forecast will continue to hold, especially when its validity requires an increasing amount of investment in the computer and semiconductor industry. In other words, experts fear that a well-known rule of economics, the law of diminishing returns, will soon come into play and lower the RoI linked to the upholding of Moore's Law.

In the present work, the author shows how this can be avoided through proper economic policy. This is because, in general, the return on any kind of investment depends on the health of the macro economy, or the basic forces of supply and demand. If a nation's total production is in balance with total demand, goods produced by industries have a ready market and an adequate profit. However, if supply exceeds demand, there is overproduction that leads to a recession and layoffs. In that milieu, the return on investment becomes mediocre and may even be negative. All this only means that proper macroeconomic policy is needed to ensure the profitability of any project.

To my knowledge, Mr. Mulay is the only writer who has made a connection between Moore's Law and macroeconomic policy. The law has played a significant role in the vast computer revolution, but the author argues that without proper economic policies the future validity of this law is at best uncertain. Mr. Mulay's contribution to the economic and technological literature is both monumental and practical. It is an innovative approach that deserves further study and research. It will appeal to those looking for new ideas.

Dr. Ravi Batra
Professor of Economics, Southern Methodist University, Dallas

Preface

In August 2007, Jack K. Steehler wrote a review for David C. Brock's book *Understanding Moore's Law—Four Decades of Innovation* in the *Journal of Chemical Education.* In his review, Jack mentioned that the book does not address a tight definition of Moore's Law. From its original article published by Intel co-founder Gordon Moore, "Cramming more components onto integrated circuits" in *Electronics Magazine* (April 19, 1965) to more recent updates, there has been a discussion focussing on many different aspects of the semiconductor industry and its productivity, rather than focusing on the fundamental definition of Moore's law. Jack believes that, given the wide ranging uses that the general concept has seen in the last 40 years, the lack of a tight focus on one specific formulation of the law is most appropriate.

The capabilities (processing speed, memory capacity, sensors) of many digital electronic devices have been improving at roughly exponential rates and are, thereby, strongly linked to Moore's Law. This exponential technological improvement in electronic devices has dramatically enhanced the impact of digital electronics in nearly every segment of the world economy. Indeed, Moore's Law has been behind the technological advancements and socio-economic developments in the late 20th and early 21st centuries.

Moore's Law has profound implications both for technology and the U.S. national as well as global economies. As long as it can be sustained, we can continue benefiting from the technological innovations and newly advanced consumer electronic goods. While Moore's Law is, on the one hand, a law of physics, many semiconductor industry professionals believe that the economics of manufacturing—specifically, the high costs of investment in shrinking transistor dimensions—will force its premature end. That would be disastrous for the economy: the end of growth for a huge sector and associated sectors that depend upon it.

While Moore's Law progressed predictably on the physical side as transistor dimensions shrank, macroeconomics was completely ignored by American businesses. Over the past 50 years, Moore's Law has been scaling at all costs and ignoring macro-economy. Moore's Law can easily continue for the foreseeable future if the chip manufacturing industry becomes sustainable by having a balanced economy. That will require some major macro-economic reforms for eliminating the gap between supply and demand caused by the gap between wages and productivity. Restoring a free-market economy in the U.S. will not only ensure the progress of high technology and innovation, thereby sustaining the progress of Moore's Law, but will also help the global semiconductor industry progress to 450 mm diameter silicon wafers in order to improve its profitability from mass production.

Macroeconomic reforms have become critical for the progress of Moore's Law as well as for the transition to 450mm diameter silicon wafers to ensure that money does not remain inert and keeps circulating in the economy. This circulation of money is critical to maintaining increasing consumer demand for the latest and greatest electronic products. Without macroeconomic reforms, the progress of Moore's Law seems impossible and the chances of the U.S. economy transitioning from the present great recession into an economic depression seems inevitable. The sustainable progress of Moore's Law advocates the "heresy" of transforming the current economic orthodoxy of monopoly capitalism into free-market capitalism. This solution involves minimal government intervention, but still requires new economic policies and business models that can help revive the U.S. microelectronics industry.

The human race has benefited exceptionally from the Internet revolution. In the last few years, there has been a dip in PC sales and a multifold increase in smartphone sales. With the high number of smartphones connected on networks comes big data as well as mobile traffic, which has also increased almost eightfold in the last few years. Moving forward, the next big thing after the mobile revolution that everybody is looking forward to is the "Internet of Things" (IoT). As the name suggests, it's a network of things, wherein things can be any smart device (like smartphones, smart home thermostats, blood pressure monitors, industrial sensors, network-connected cameras, etc.) that can communicate with other smart devices. While some analysts believe that the IoT market would achieve 5.4 billion connections worldwide by 2020, up from 1.2 billion devices today, the poor consumer purchasing power in the global economy makes this analysis truly questionable.

Sustaining Moore's Law takes the reader through this journey of the progress of Moore's Law and explains how this technological progress has become unsustainable due to the violation of macroeconomic parameters of economy. Along with offering solutions to sustain the progress of Moore's Law, this book also provides a blueprint for a sustainable increase in the profitability of the global semiconductor industry in this age of falling growth by means of transitioning to 450 mm wafers. It also engages with the semiconductor industry thought leaders about the future of the global semiconductor industry and the IoT revolution.

Chapter 1 exposes the reader to the importance of Moore's Law and the resulting human progress. Chapter 2 highlights a feasible path for transitioning from an unsustainable to a sustainable progress of Moore's Law. Chapter 3 helps the reader understand the impacts of semiconductor business models on the national interests of a country. Chapter 4 provides an analysis of the impacts of semiconductor business models on the sustainability of this industry. Chapter 5 deals with the macroeconomic cycles and proposes modifications to existing fabless-foundry business models in the semiconductor industry for the progress of Moore's Law. Chapter 6 forecasts the demise of Moore's Law due to physics rather than economics. Chapter 7 evaluates the importance of a decentralized supply chain in ensuring the sustainability of the IoT revolution.

Chapter 8 offers macroeconomic solutions for transitioning to 450 mm diameter wafers for the semiconductor industry. Chapter 9 provides a detailed analysis of the progress of Moore's Law and offers a sustainable path for advancing this law beyond its 50th anniversary. Chapter 10 talks about the macroeconomics of semiconductor manufacturing in a developing economy like

India's. Chapter 11 offers macroeconomic solutions for ensuring the success of the IoT revolution. Chapter 12 engages with the thought leaders of the fabless semiconductor industry in the U.S. and offers macroeconomic solutions for ensuring the sustainable progress of Moore's Law for the success of the IoT revolution.

Apek Mulay
September 2015

Acknowledgments

I owe my greatest intellectual debt to the great neohumanist *Shrii Prabhat Ranjan Sarkar,* whose ideas about neohumanism and human progress continue to inspire me to do my part for the greater good of humanity. I am very thankful to Morgan & Claypool for bearing the costs of publishing this book. I should also express my gratitude to *Electronics.ca Publications*, Truthout.org, EBN, *Military & Aerospace Electronics, The Economic Times,* LinkedIn, and Semi.org for their permission to let me use my published work in this book. I would also like to extend my thanks to semiconductor industry professionals around the world who continue giving their best for the continued growth of this industry and who benefit mankind with their technological innovations. I would also like to thank Dr. Kris Iniewski for his support in this venture.

Above all, I am thankful to all my family members and friends for their love and support without which it would not have been possible for me to make this achievement.

Apek Mulay
September 2015

CHAPTER 1

Impacts of Moore's Law on Human Progress

1.1 INTRODUCTION

Human beings have been constantly progressing physically and intellectually, overcoming all the obstacles in their path. Human beings continue their endless efforts for progress because to remain stagnant is contrary to human wants. The obstacles in the path of progress have always resulted in clashes and cohesion. Every progression has occurred through these clashes and cohesion. Today, we observe that human beings are racking their brains in the search for new ways to win battles, and therefore their brain cells are also developing. They are devising new ways to preserve past history. After thinking for some time humans devised words as pictures. By looking at the pictures, they remembered what they did a few days ago. This is called pictography. The pictorial script was devised in this way. Even today, China uses a pictorial script—the alphabet in pictures.

Human beings developed all of this out of necessity. The quest to store, retrieve, and process information is one task that makes humans different from other animals. Thus, the human civilization evolved out of its primitive past. No known animal uses tools to store, retrieve, and process information. Moreover, the social and technological progress of the human race can be directly traced to this human attribute.

1.2 THE HUMAN MIND AND DATA STORAGE

Thinking and memorizing are two functions of the human mind. The more the thinking capacity increases, the more the power of memory develops. The nerve cells also change, leading to a corresponding change in the nerve fibers. These changes create a stir and a revolution in the world of thought. No other creature thinks like this—only humans have the power of contemplation. Human literature, art, etc. reflect this sort of contemplation. With the aid of modern science, the people of today do not have to labor quite as hard as our ancestors had to do in order to discover the way to progress further. This endeavor to progress further has made human beings superior beings.

Man's earliest attempts to store, retrieve, and process information date back to prehistoric times when humans first carved images in stone walls. Then in ancient times, Sumerian clay tokens developed as a way to track purchases and assets. By 3000 B.C., this early accounting tool had developed into the first complete system of writing on clay tablets. Ironically, these were the first silicon-based storage technologies and would be abandoned by 2000 B.C., when the Egyptians developed papyrus-based writing materials. It would take almost four millennia before silicon would stage a comeback as the base material, with the main addition being the ability to process stored information. In 105 A.D., a Chinese court official named Ts'ai Lun invented wood-based paper. But it wasn't until Johann Gutenberg invented the movable-type printing press around 1436 that books could be reproduced cost effectively in volume.

1.3   EARLIER FORMS OF STORAGE MECHANISM

The first large book that was published by Johann Gutenberg was the Gutenberg Bible, published in 1456. Something akin to Moore's Law occurred, as Gutenberg went from printing single pages to entire books in 20 years. At the same time, resolution also improved, allowing finer type as well as image storage. Yet, this was primarily a storage mechanism.

It would take at least another 400 years before retrieval would be an issue. In 1876, Melvil Dewey published his classification system that enabled libraries to store and retrieve all the books that were being made by that time. Alan Turing's "Turing Machine," first described in 1936, was the step that would make the transformation from books to computers. So Moore's Law can be seen to have a social significance that reaches back more than five millennia.

1.4   ECONOMIC VALUE OF MOORE'S LAW

The economic value of Moore's Law is also understated, because it has been a powerful deflationary force in the world's macro-economy. Inflation is a measure of price changes without any qualitative change. So, if price per function is declining, it is deflationary. Interestingly, this effect has never been accounted for in the national accounts that measure inflation-adjusted gross domestic product (GDP). The main reason is that, if it were, it would overwhelm all other economic activity. It would also cause productivity to soar far beyond even the most optimistic beliefs. This is easy to show, because we know how many devices have been manufactured over the years and what revenues have been derived from their sales.

A logical way to analyze the economic impact of Moore's Law is price per transistor. The benefits of Moore's Law towards human progress can be traced to the universal value to an end user in the form of transistors. The more the number of transistors on an electronic circuitry, the greater is the functionality of the product the consumers can buy. In this way, the number of transistors translates into system functionality. Therefore, by cramming more transistors on an integrated circuit, it is possible to not only add value to the final product but, by means of mass production, it becomes easy to reduce the costs of production. In this way, Moore's Law has been able to offer greater value to the semiconductor industry by offering a higher value to consumers and reducing costs for manufacturers. In this way, it has contributed to ever new innovations in consumer electronics and provided a profitable business model for the semiconductor industry.

The constancy of this phenomenon is so stunning that even Gordon Moore has questioned its viability. Moore's Law proved to be a predictable business model for the semiconductor manufacturing industry from its inception in 1965 through the end of the millennium, up to today. When Gordon Moore came up with his observation, he understood the economic impacts of the progression in transistor technology nodes. In his assessment, Moore also took into consideration user benefits, technology trends, and the economics of manufacturing. Moore had confidence in the future of the semiconductor industry because it looked predictable. The planning and investing components of the business model were based on Moore's Law, which states that the integration scale of transistors would always rise within a year or two. This would make the existing electronics in the market obsolete, and the faster and cheaper new products that would be introduced with the evolution of transistor technology nodes would help generate greater demand for those products. Economic incentives encourage producers to crowd more and more components on each integrated circuit, which proportionately increases the speed with smaller transistors and proximity of components.

1.5 WHAT DRIVES TECHNOLOGICAL INNOVATIONS

The human psyche demands at different ages drives the technological innovations of that age. In order to overcome physical and psychic problems and inconveniences, the people of a particular era invented and popularized bullock carts for transportation. Later they developed faster horse-drawn carriages. Subsequently, as time passed, public demand also changed. That is why different types of transportation, such as motor cars, aeroplanes, rockets, etc. have been invented at different times. None of these inventions should be condemned. They are all simply designed to meet the demands of different ages.

Conflicts in the physical sphere gradually awaken dormant human potential. Environmental influences also increase the degree of complexity of the human body. The problems of ancient and modern people cannot be considered as identical. To keep pace with the changing problems of life, the human body and mind have gradually become more complicated. The physical structures of ancient humans would have certainly been unfit for solving the problems of today. As the mind becomes more complex, its direct centers, the nerve cells, and its indirect centers, the glands, undergo corresponding biological changes. As the nature of problems change, the human mind responds by making new scientific discoveries and technological innovations. Hence, a steady cultivation of science and technology must go on and such cultivation will never be an impediment to human progress.

1.6 CONCLUSION

The progress of Moore's Law has made significant contributions to economic and human advancement. There has been a rapid growth in technological progress because of the predictable business model based on the continued progress of Moore's Law, which ensures steady profitability and productivity. The continued progress of humanity depends on the scientific and technological progress of Moore's Law for the benefit of humanity.

1.7   REFERENCES

[1] Hutcheson, Dan G., "Moore's Law: The History and Economics of an Observation that Changed the World," *Electrochemical Society Interface*. Spring 2005, pp. 17–21.

[2] Moore, Gordon E., "Lithography and the Future of Moore's Law," *SPIE Speech*. 1995.

[3] Mulay, Apek, *Mass Capitalism: A Blueprint for Economic Revival*, Book Publishers Network. Bothell, WA, 2014.

[4] Sarkar, P. R., *PROUT in a Nutshell*, Ananda Marga Publications, Kolkata, India. 1959.

CHAPTER 2

From an Unsustainable to a Sustainable Progress of Moore's Law

2.1 INTRODUCTION

The price or cost per transistor has been used as one metric of measuring the economic impacts of Moore's Law. The cost per transistor is an exceptionally good metric because it can easily be translated into a universal measure of value to a user: transistors. Transistors are a good measure because in economic terms they translate directly into system functionality. The more the number of transistors, the greater the functionality of electronic products that consumers can buy. Hence, a lower cost per transistor translates into a higher buying power for the consumer. Hence, the goal of the progress of Moore's Law is essentially growth in consumer purchasing power for the end user.

2.2 UNSUSTAINABLE PROGRESS OF MOORE'S LAW–A PROGRESS OF "SUPPLY-SIDE ECONOMICS"

In recent times, there is a growing concern about the state of the global semiconductor ecosystem. The entire economic structure that was supposed to lead to next-generation manufacturing technologies, like 450 mm wafers, extreme ultraviolet (EUV) lithography, and transistor technology nodes below 14 nm CMOS, is on the verge of coming apart due to poor consumer demand for electronic goods. Whether it is transition to 450 mm wafers, higher power requirements for EUV lithography, or shrinking transistor dimensions for keeping up with progress of Moore's Law, the problems facing the semiconductor ecosystem are essentially the same—increased manufacturing costs that result in poor RoI threaten our progress.

Whether you take into consideration 450 mm silicon wafers, EUV lithography improvements, or scaling transistor geometries, from an economic standpoint there is one thing that is common to all their respective contributions to the supply of silicon to the economy. While 450 mm silicon wafers would increase the diameter of wafers for mass production to reduce the costs, the increased yield from larger wafer sizes contributes to an increased supply of silicon. When it comes to EUV, no one, including lithography systems maker ASML, has yet demonstrated an EUV tool capable of providing the necessary source power concentrations for sustaining production volumes.

This evolution of EUV lithography would again make a contribution only to the supply of silicon wafers. The relentless progress of Moore's Law, since Gordon Moore made his famous observation in 1965, has essentially been reducing manufacturing costs by scaling geometries, thereby reducing prices for consumers while providing silicon with a higher performance. The progress of Moore's Law has essentially been the progress of supply-side economics. There has been little to no incentive to boost consumer demand in the U.S., as well as global, economy besides luring consumers into an unsustainable debt. The term "supply-side economics" was thought, for some time, to have been coined by journalist Jude Wanniski in 1975, but this term "supply-side" ("supply-side fiscalists") was first used by Herbert Stein, a former economic adviser to President Nixon, in 1976, and only later that year was repeated by Jude Wanniski.

Its use connotes the ideas of economists Robert Mundell and Arthur Laffer. Supply-side economics is likened by critics to "trickle-down economics," a theory that believes money trickles down from the producers to the consumers. Trickle-down economics advocates that producers are job creators in an economy and, hence, it supports giving tax cuts to the producers as a way to boost economic growth so that producers can hire more employees. However, there is a flaw in this approach because, if the producer is not able to sell what he/she has already produced, why would the producer consider hiring more employees (even with an added incentive of tax cuts) if the consumer demand for his product does not increase?

In an interesting article, "Reagan: The Great American Socialist," published on Truthout.org, economist Ravi Batra argues that the supply-side economic policies adopted by Ronald Reagan's economic advisors in 1981 resulted in budget deficits soaring from 2.5% of GDP to more than 6% of GDP, alarming financial markets, sending interest rates sky-high, and culminating in the worst recession since the 1930s. Reagan's supply-side economic policies resulted in the wealthiest facing a 28% tax rate, while those with lower incomes faced a 33% rate. In addition, the bottom rate climbed from 11% to 15%. This is how supply-side economic policies since their inception have increased the income disparity across U.S., as well as global, economy. This is verily why free-market capitalism has been transformed into crony capitalism over the years, which is now bringing this global economy and global semiconductor industry to a standstill due to poor consumer demand resulting from a poor RoIs.

2.3   FREE-MARKET ECONOMY–A PATH TOWARDS SUSTAINABLE PROGRESS OF MOORE'S LAW

We need to implement economic solutions for sustaining the progress of Moore's Law through establishment of a true free-market economy where the real job creators in the economy are not only producers but also consumers. Without a healthy consumer demand for the latest and greatest electronic products, any further investments in the progress of Moore's Law—transitions to 450 mm silicon wafers and EUV lithography improvements—are bound to provide a poor RoIs for the producers.

If free-market economic reforms can usher in a healthy consumer demand for electronic products, then the RoIs for the producers can also be ensured. Hence, we can conclude that the slowing of EUV rollout, halting of 450 mm wafers, and uncertainty in economic demand resulting in poor RoIs beyond 14 nm technology node is essentially a failure of supply-side economics and could not be claimed to be the demise of Moore's Law.

The minimum necessities of human society should be met through a growth in consumer purchasing power in the economy. Semiconductor industry professionals are not only to recognize the importance of higher consumer purchasing power in the economy but should also be actively involved in ushering true free-market economic reforms that guarantee the growth in consumer purchasing power in the economy for the continued progress of Moore's Law. It also becomes a social responsibility to provide individuals with higher purchasing power. When such free-market reforms based on mass capitalism are carried out, there would be special incentives provided to individuals with special abilities. Each and every human being requires clothes, medicine, housing accommodation, proper education, food for proper nourishment, etc. These demands must be fulfilled by means of providing work by creating jobs and not by means of offering any kind of dole.

While free markets ensure that wages catch up with productivity, there should be special amenities provided to intellectuals, scientists, and people performing special services. However, when offering these amenities, it needs to be ensured that the consumer purchasing power of the overall economy also grows, while minimizing the growth of huge disparities in the economy. There will always be a gap between the minimum necessities, which decide the standard of living, and special amenities that are offered to meritorious people. A free-market approach to minimize the gap between minimum necessities and maximum amenities is to raise the consumer purchasing power through a productive use of technology, which would also boost the minimum amenities for all. For example, in a certain country, leaders and intellectuals would require a luxury car, and these should be offered to them because of their profound contribution to the society. But after that endeavor, the buying power of other individuals should also proportionally increase so that they are able to afford to buy at least a motor bike, if not a car.

However, in this process of improving the consumer purchasing power by productive use of technology, it could be noticed that, after some time, the purchasing power of leaders and intellectuals needs to be higher for them to be able to afford an airplane. In that regard, the consumer purchasing power of the rest of economy (excluding the leaders and intellectuals) should also be high enough (to keep economic disparity in check) so that they are be able to afford at least an ordinary car, if not a luxury one. Diversity is the inevitable law of nature. However, if this approach is taken, then it would be possible to raise the standard of living for all citizens in an economy while still rewarding hard work and merit. In fact, this approach would be a sustainable one, as it would ensure that the consumer debt and national debt do not grow due to the growth in consumer purchasing power of the overall economy, thereby minimizing huge economic disparity. Although the gap between minimum necessities deciding the standard of living of citizens and the maximum amenities that are provided to meritorious people will remain unbridged forever, this gap should not exceed the certain limits that have resulted in recessions and depressions in the global economy. In this way, a productive use of technology would increase the consumer purchasing power such that wages will keep track with productivity, thereby sustaining the progress of Moore's Law through growth in consumer purchasing power. One example of productive use of technology over an unproductive use of the same can be explained based on unemployment created due to automation.

Today, many modern economic thinkers blame automation, based on technological progress, as a major cause of job losses. However, technology could be productively utilized in such a way that the manufacturing sector could cut back on work hours while paying workers a high wage due to their high productivity. This is because automation enables a worker to be highly productive through the use of machines to manufacture products. High worker productivity significantly increases the supply of goods to the economy. As a result of this increased productivity, workers would be able to work for a fewer number of hours to achieve their production target. They could use their spare time to pursue higher education, leisure, hobbies, vocational training, etc. In this way it is also possible to minimize, if not eliminate, the problem of high unemployment resulting from automation, while still keeping the supply of goods proportionate to consumer demand, thereby maintaining an economic balance. This is how productive utilization of technology would not lead to a problem of unemployment.

2.4   CONCLUSION

The progress of Moore's Law since its inception has been the evolution of supply-side economics. However, this has resulted in not only an increase in the supply of goods to the economy, but has also resulted in a very poor consumer demand for these goods. For a sustainable progress of Moore's Law, not only the supply but also the demand has to increase proportionally, leading to balanced economic growth.

2.5   REFERENCES

[1] Batra, Ravi, "Reagan: The Great American Socialist," Truthout.org. March 20, 2009.   http://www.truth-out.org/archive/item/83147:reagan-the-great-american-socialist

[2] Hruska, Joel, "This is what the death of Moore's Law looks like: EUV rollout slowed, 450 mm wafers halted, and an uncertain path beyond 14 nm," *ExtremeTech.*, March 17, 2014. http://www.extremetech.com/computing/178529-this-is-what-the-death-of-moores-law-looks-like-euv-paused-indefinitely-450mm-wafers-halted-and-no-path-beyond-14nm

[3] Mulay, Apek, *Mass Capitalism: A Blueprint for Economic Revival,* Book Publishers Network. Bothell, WA, 2014.

[4] Sarkar, P. R, *PROUT in a Nutshell,* Ananda Marga Publications, Kolkata, India. 1959.

[5] Wikipedia, "Supply-side economics." `http://en.wikipedia.org/wiki/Supply-side_economics`

CHAPTER 3

Impacts of Semiconductor Business Models on U.S. National Interests

3.1 INTRODUCTION

The business models[1] prevalent in the semiconductor industry have had a profound impact on the progress of Moore's Law. The semiconductor industry has undergone a transformation from Integrated Device Manufacturers (IDMs) business model to Fabless-Foundry business model. The fabless-foundry business model has resulted in increased collaboration between fabless semiconductor companies and their foundries. However, the policies of globalization led to the transformation of the entire U.S. Semiconductor Industry from a few integrated device manufacturers (IDMs) to several fabless small businesses, leading to new innovations in the microelectronics business. Deceptive "free trade" agreements have resulted in not just a transfer of manufacturing technology to China, but also in increased threats of counterfeit electronics entering the U.S. supply chain. This chapter explains how U.S. National security may be impacted by the transfer of U.S. semiconductor manufacturing technology.

3.2 ECONOMIC REFORMS CAN DEFEND AGAINST COUNTERFEIT ICS

In addition to nearly $600 billion in trade deficits that have resulted from free-trade policies, the counterfeit electronics entering the United States supply chain from China has become a national security threat for the United States. Initially, the U.S. manufactured all defense-related products at home, while building consumer electronics in China to capitalize on low labor costs of manufacturing. The progression to advanced transistor technology, however, required increasingly large investments on the part of private defense contractors that manufactured semiconductor wafers in the U.S.

Hence, several independent and for-profit defense contractors based in China started making use of Chinese-built ICs for military weapons such as missiles, machine guns, and drones.

---

[1]The first version of this text appeared in Truthout.org, http://truth-out.org/speakout/item/22910-transformation-of-us-semiconductor-industry. Reprinted with permission.

The state-of-the-art infrastructure and technical know-how to make advanced technology products have also been transferred to China. China is now flooding the U.S. defense supply chain with counterfeit ICs. The Government Accountability Office (GAO) claims that 40% of the U.S. Department of Defense's supply chain is undermined by fake or defective parts. It has become very expensive to mitigate the introduction of counterfeits into the U.S. supply chain, and this is cutting into the profits of U.S. defense contractors.

There is no solution for the twin (budget and trade) deficits under the exploitative free trade scenario. The U.S. cannot afford to run year-over-year budget deficits. To protect its sovereignty, it needs to reexamine its internal economic policy, as well as its foreign policy. We need to restore the U.S. manufacturing supply chain to ensure a sustainable economy, eliminate budget and trade deficits, balance consumer demand with supply, and minimize the wealth concentration in the economy in order to ensure more circulation of currency by means of ushering true free-market reforms based on the theory of *Mass Capitalism*.

This would ensure that wages of hard-working Americans catches up with their productivity. It would establish a free-market economy, thereby ensuring that supply and demand rise and fall with minimal government intervention. Additionally, such a system would establish a balanced economy without running any trade and budget deficits.

3.3   UNSUSTAINABLE BUSINESS MODELS OF THE U.S. SEMICONDUCTOR INDUSTRY

The growth of the fabless semiconductor industry in the last two decades—which has resulted in more silicon consumption for the manufacture of handheld mobile devices as compared to traditional PCs—has an important role to play in ongoing mobile revolution. The recent shift in consumer demand for smartphones over PCs has reduced consumer demand for Intel's processors. Intel's decision to enter into the wireless business was late, as fabless businesses like Qualcomm, Broadcom, etc. had developed a competitive edge over Intel due to matured manufacturing processes at external fabs like Taiwan Semiconductor Manufacturing Corporation (TSMC). This forced Intel to outsource the manufacturing of its wireless business unit to Asian fabs like TSMC.

Today, Intel Inc. has three idle fabs in Arizona, Oregon, and Ireland. These idle fabs have exposed the failure of the IDM business model, as idle time often indicates a loss of revenue due to accelerated depreciation of state-of-the-art tools. Intel has been forced to open up its foundry business to offer manufacturing services for fabless companies in the U.S. and Intel Inc. competes in its business with some of these fabless companies for whom it has opened up its manufacturing services. The flaw in the existing fabless business model of U.S. companies lies in U.S. trade policies. "Free trade" policies have enabled Asian fabs to manufacture and export goods back into the U.S. without any kind of import duty. This fabless-foundry business model is a major contributing factor to U.S. trade deficits.

3.4 ADVANTAGES OF AN INDEPENDENT DOMESTIC FOUNDRY TO THE U.S. FABLESS SEMICONDUCTOR INDUSTRY

There are several advantages to the business model whereby U.S. fabless businesses get manufacturing done from an independent domestic foundry. First, the U.S. is running a huge annual trade deficit with countries like China and Japan as a result of sending semiconductor manufacturing to these countries. Hence, when manufacturing gets done domestically, the trade deficits would be eliminated over time. Second, free trade policies with Asian countries like China have also resulted in a distorted supply chain of manufacturing. The centralized supply chain that is followed by U.S.-based MNCs to increase their annual profits has resulted in rising trade deficits, which have resulted in a negative economic growth in U.S. economy for the first quarter 2015 and an economic stagnation in subsequent quarters. However, these "free trade" practices have also resulted in the introduction of counterfeit electronics into the U.S. supply chain and it has become very expensive to mitigate this. Third, by creating more jobs domestically, U.S. unemployment could be trimmed and the government would not have to spend money to pay for the unemployment benefits of laid-off workers.

3.5 UNITED ARAB EMIRATES' (UAE'S) TRACK RECORDS ON TECHNOLOGY RE-EXPORT

The UAE record on preventing re-export of advanced technology, particularly to Iran, has been mixed in the past few years. Taking advantage of geographic proximity and the high volume of Iran-Dubai trade ($10 billion per year), numerous Iranian entities involved in Iran's energy sector have offices in the UAE that are used to try to procure much-needed technology and equipment. The Institute for Science and International Security issued a report in January 2009 entitled "Iranian Entities' Illicit Military Procurement Networks," which asserted that Iran has used UAE companies to obtain technology from U.S. suppliers and that the components obtained have been used to construct improvised explosive devices (IEDs) shipped by Iran to militants in Iraq and Afghanistan. The report also alleged that other UAE companies that were involved in this network included not only Mayrow Electronics, but also Majidco Micro Electronics, Micatic General Trading, and Talinx Electronics, which are headquartered in Iran.

3.6 U.S.–UAE TRADE DEAL

According to the statistics from the U.S. Census Bureau's Foreign Trade Division, the UAE has been the largest market for U.S. exports to the Middle East for four years in a row. In 2012, U.S. firms exported nearly $24.81 billion worth of goods to the UAE. In addition to ongoing free trade talks between the U.S. and UAE, as part of the Gulf Co-operation Council (GCC), the UAE is negotiating with the United States a "GCC-U.S. Framework Agreement on Trade, Economic,

Investment, and Technical Cooperation," an umbrella instrument for promoting ties between the two sides in the economic area—essentially a GCC-wide trade and investment framework agreement (TIFA). Since GlobalFoundries Inc. has a 300 mm semiconductor foundry planned in Abu Dhabi in the future, it is obvious that, because UAE's ATIC (Advanced Technology Investment Corporation) owns the GlobalFoundries Fab in Malta, New York; it would use this ownership as a leverage to transfer manufacturing technology from its fab in New York to its upcoming fab in Abu Dhabi. If manufacturing technology does get transferred from the New York fab to the Abu Dhabi fab in the future, Iranian scientists and engineers working in Abu Dhabi could easily get access to this manufacturing technology, due to geographic proximity and the high-volume trade deal between Iran and Dubai. In the past, free trade agreements with China have resulted in the introduction of counterfeit electronics into the U.S. supply chain and ownership of trillions of USD by China in its foreign exchange (forex) reserves.

3.7   CONCLUSION

In order to avoid the transfer of advanced semiconductor manufacturing technology to politically unstable regions of the world like Iran, which have supported radical Islamist and terrorist groups in the Middle East, the U.S. semiconductor industry should have its manufacturing done in a domestic foundry and should not rely on any kind of foreign investments. Therefore, the top notch foundries in the U.S. supporting the fabless semiconductor businesses in the U.S. should find a way to become independent of their foreign investors. If domestic fabs in the U.S. decide to go for an initial public offering (IPO), based on the history of Wall Street when it comes to a capital intensive business, Wall Street investors would pressurize shareholders of these fabs to ship manufacturing abroad due to the high cost of upkeep, maintenance, and upgrade of the wafer fab. The other option is to partner with the US government. A symbiotic partnership with the government would make this foundry business very sustainable.

Only with the backing of the U.S. government can any top-notch wafer fab be at the leading edge of technology development with sustainable capital investments, helping to keep the foundry independent of any foreign capital. Additionally, ownership of a top-notch fab by the U.S. government, as compared to any foreign government, would also give confidence to the U.S. fabless semiconductor industry, due to protected intellectual property for those fabless businesses. To keep this partnership with the U.S. government symbiotic and efficient, the business model of the U.S. semiconductor industry should transform from a globalization-based model to a sustainable business model. One such model will be presented in Section 5.3, "A Three-Tier Business Model for the Semiconductor Industry." This new business model would transform the entire U.S. semiconductor industry and generate high profits for years to come by means of a growth in consumer purchasing power in the economy. Such a transformation of industry would take the US Semiconductor industry to the next level of innovation and financial success.

3.8   REFERENCES

[1] Kenneth Katzman, "The United Arab Emirates (UAE): Issues for U.S. policy," Congressional Research Service, May 15, 2014.

[2] Mulay, Apek, *Mass Capitalism: A Blueprint for Economic Revival,* Book Publishers Network, Bothell, WA, 2014.

[3] Mulay, Apek, "The state of the global economy," May 13, 2015. `http://www.truth-out.org/speakout/item/30774-the-state-of-the-global-economy`

[4] U.S.–UAE Business Council, "Record-breaking U.S. export numbers showcase vibrant U.S.–U.A.E. trade growth," February 8, 2013. `http://usuaebusiness.org/2013/02/record-breaking-u-s-export-numbers-showcase-vibrant-u-s-u-a-e-trade-growth/`

CHAPTER 4

Impacts of Semiconductor Business Models on Sustainability

4.1 INTRODUCTION

Two prevalent business models in today's global semiconductor industry are the integrated device manufacturer (IDM) model and the fabless-foundry model. In this chapter, we analyze the superiority of the fabless-foundry business model over IDM model. However, this chapter also analyzes the importance of complying with the common-sense macroeconomic parameters of economy for a long-term sustainability of the fabless-foundry business model.

4.2 THE TRANSFORMATION OF THE U.S. SEMICONDUCTOR INDUSTRY BUSINESS MODEL

Today, the U.S. semiconductor industry is going through a very interesting phase of transformation. During the past couple of decades, the growth of the fabless semiconductor industry has played a dominant role in the transformation of the U.S. semiconductor industry. Today, the majority of the IDMs, like Texas Instruments, Advanced Micro Devices, Motorola, etc., have either spun off their manufacturing facility to become fabless or have shipped their advanced semiconductor manufacturing operations to Asian fabs. This transformation occurs because the silicon business has become incredibly expensive for integrated-device manufacturers in the U.S., which have to invest huge amounts of money in R&D to develop the next generation of silicon technology.

The same costs are also involved in the development of process technology by the independent foundries that manufacture for the fabless semiconductor companies. Such foundries manufacture for several fabless companies. They have a significantly higher consumer demand for their silicon, compared to IDMs, because foundries for fabless companies cater to a larger customer base. Today, a fabless semiconductor company can eliminate the costs involved in the development of next-generation technology and invest in the design of new innovative products. Pure-play foundries catering to these fabless companies can cater to a much larger customer base with their mature manufacturing processes. Therefore, the fabless-foundry semiconductor ecosystem has turned out to be a win-win situation for both the fabless semiconductor companies and their foundries. This mutually profitable business model, based on co-operative collaboration, has been a success story of the U.S. fabless-foundry semiconductor ecosystem.

Due to outsourcing the huge investments in the manufacture of ICs, fabless businesses have additional revenue to focus on new circuit designs and thereby are able to provide many more innovative solutions to the US semiconductor industry. On the other hand, due to ever-increasing investments in the next-generation silicon-process development, IDMs have become cash-strapped for their investments in innovative designs. As a result of fewer innovations in their products, IDMs also have a lower consumer demand for their products. Due to this lower consumer demand, the wafer fabs of IDMs remain idle to avoid any overproduction of silicon. Fab idle time is also one of the factors that acts as a drain on the annual revenue of a foundry because the wafer fab tools are incredibly expensive for next-generation process development. Also, for next-generation silicon-process development, there is an accelerated depreciation of tools, which permits the industry to progress to a new silicon technology node. Hence, tool-idle time for a fab is extremely expensive, as technology progresses to sub-nanometer technology nodes for the progress of Moore's Law.

IDMs convince the state and local governments that IDMs are creating employment through huge investments in next-generation wafer fabs and they look for tax incentives from governments in order to make these capital-intensive investments in the economy. These tax incentives also act as a drain on the economy because the tax cuts provided to IDMs significantly reduce government's revenue for performing its day-to-day business and result in budget deficits if IDMs do not create a sufficient number of jobs in the economy. Additionally, if IDMs build up top-notch wafer fabs because of government-offered tax incentives but are not able to sell all wafers that they have produced, then the IDMs have to lay off their employees. The recent layoffs at semiconductor IDMs like Intel Inc. are evidence of the failure of such economic policies. One more reason that the IDM business model is inferior when compared to the fabless-foundry semiconductor business model is that fabless companies produce more varieties of innovative products because of diversity in their design engineering teams. Also, the foundries that cater to these fabless businesses expand their capacity when they have more customers. Expanding a business based on growing the customer base is certainly a much wiser and sustainable business decision compared to expanding a business just because of tax incentives received from the government.

Thus, I have demonstrated the economics behind the success of the fabless-foundry semiconductor business model over the IDM business model. If the U.S. semiconductor industry is to keep on track with Moore's Law, it must opt for the fabless-foundry semiconductor business model over the IDM model. Hence, the future is bright for the growth of the fabless-foundry semiconductor business model while the IDM model is sure to fall behind. It is just a matter of time before the IDM model will not be able to keep up with the requirements of huge capital investments for the next generation of semiconductor manufacturing to sustain the progress of Moore's Law. Although the future of the US semiconductor industry seems bright with the fabless-foundry semiconductor business model, the demise of Moore's Law could happen much earlier because of trade policies and other macroeconomic policies that have made the present progress of this industry unsustainable. Hence, the existing fabless-foundry business model must undergo reforms so that macroeconomic losses to the U.S. economy due to the globalization of semiconductor manufacturing are eliminated.

4.3   MACROECONOMIC CYCLES AND BUSINESS MODELS

Macroeconomics and macroeconomic conditions play a critical role in the operation of any business. Entrepreneurs and others business people must take these factors into consideration as part of their market analysis. As one example, the reduction or increase in demand for products affects decisions to expand or scale down their rate of production. Macroeconomics is intertwined with business because business is affected by the factors that constitute macroeconomics. Macroeconomics deals with issues relating to factors that affect the economy, including areas like the rate of unemployment, inflation, business cycles, and gross domestic product (GDP).

In this section, I explain the role of macroeconomic cycles and how they relate to business models prevalent in the semiconductor industry. The semiconductor industry in the United States has undergone a transformation from a few IDMs to several fabless semiconductor companies. The fabless-foundry business model has led to a transformation of the U.S. semiconductor industry. However, the fabless-foundry business model violated commonsense macroeconomic parameters of the U.S. economy, hence, the macroeconomic losses to the U.S. economy began to outweigh the microeconomic benefits from this approach. The U.S. semiconductor industry began offshoring labor-intensive manufacturing operations in the 1960s, followed in the 1970s and 1980s by increased offshoring of complex operations, including wafer fabrication and some research and development (R&D) and design work. Although offshoring of labor-intensive semiconductor jobs from the U.S. to Japan had started way back in 1960s, the macroeconomic losses then were minimal, due to prevalent free-market capitalism in the U.S.

When the Federal Election Campaign Act (FECA) was passed in 1971, the resulting political corruption due to unlimited contributions of "soft money" transformed free-market capitalism in the U.S. into crony capitalism. The budget deficits started to soar when huge tax cuts were passed under Ronald Reagan's presidency. The graph in Figure 4.1 below shows that, since the FECA was passed in 1971, the monetary policy started to change due to political corruption so that real wages of employees started to trail employee productivity.

This monetary policy resulted in a increasing gap between wages and productivity. The tax cuts passed by President Reagan resulted in a growing disparity in the United States. Figure 4.2 shows a comparison between the Great Depression and Great Recession since 2007 as a result of growing economic disparity, showing clear evidence that a great depression has actually started in the U.S. in 2008. The disparity started to grow in 1980 with Reagan's tax cuts for the wealthy and, when this economic disparity came close to what it was just before 1929 Great Depression, stock markets crashed in 2008.

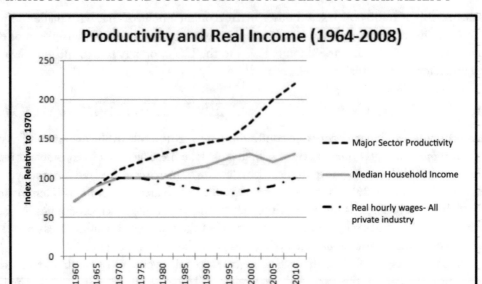

Figure 4.1: Productivity and real income (1964–2008) when indexed to 1970. As shown in this figure, the productivity of American workers has steadily increased but their wages have trailed their productivity.

4.4   MACROECONOMIC FORECASTS FOR THE U.S. ECONOMY BASED ON MACROECONOMIC CYCLES

Let us analyze the macroeconomic cycles that have resulted in the Great Recession of 2007. Everything governed by the laws of nature moves in a systaltic fashion and never in a straight line. Due to this systaltic motion, internal clash and cohesion take place, giving rise to economic cycles. The ups and downs of socio-economic life in different phases are sure to take place due to this systaltic principle.

These macroeconomic cycles generally last 7–14 years. But, when the commonsense macroeconomic parameters are grossly violated, the macroeconomic cycle shifts to a shorter period and in the 7th year of the cycle, the economy starts losing its momentum. Since the tax cuts in 1980, budget deficits started to soar and trade deficits were already increasing as a result of free-trade policies since World War II. In fact, the fabless-foundry business model has resulted in increased offshoring of even the capital-intensive jobs along with ongoing offshoring labor-intensive jobs from the United States due to free-trade policies of the United States.

The macroeconomic parameters that were violated since 1980 translated into an increasing gap between wages and productivity. The U.S. has sustained these deficits by failing to fix the

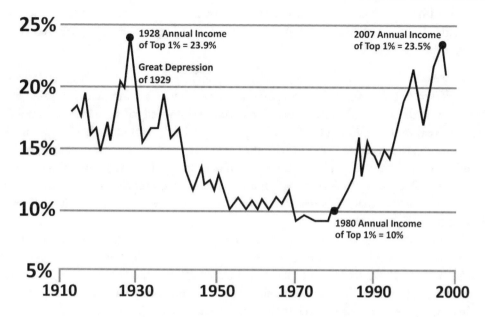

Figure 4.2: Annual income of the richest 1% of the population from 1910–2010. When the disparity has increased such that just 1% earns close to 24% of the annual income in the economy, the economy has gone into a depression.

root cause of problem (i.e., the growing gap between wages and productivity) and printing USD instead. Now, let us look at the evidence of macroeconomic cycles post Reagan's tax cuts in 1980s.

1. In finance, Black Monday refers to Monday, October 19, 1987, when stock markets around the world crashed, shedding a huge value in a very short time. This occurred seven years after the tax cuts of 1980.

2. 1994 wasn't a particularly notable year. There was no presidential election, no major geopolitical developments, no stock market crash, and no summer Olympics. However, in 1994, the bond market suffered a very sharp and sudden sell-off that started in the U.S. and Japan and then spread more or less across all developed markets. This was seven years after 1987 crash.

3. On September 17, 2001, U.S. stocks plunged to the lowest levels in nearly three years and the Dow Jones industrial average suffered its worst point-loss in history when trading resumed for the first time following the September 11, 2001 terrorist attacks. This was seven years after 1994.

4. According to ex-Fed chair Ben Bernanke, September and October of 2008 was the worst financial crisis in global history, including the Great Depression. This occurred seven years after the 2001 crash.

As we can observe, the seven-year macroeconomic cycles have never been violated because of the violation of commonsense macroeconomic parameters resulting in an increased economic disparity since the 1980s. Recently, Alan Greenspan acknowledged that the economic demand today is as weak as it was during the Great Depression, which concludes what I have presented in Figure 4.2 comparing today's economy with the Great Depression. Some well-known economists, such as the Nobel Laureate and Princeton economics professor Paul Krugman, also call the slump that started in 2007 a depression instead of a great recession. Hence, based on these macroeconomic cycles, I believe that it is a near certainty that in 2015, seven years after 2008, there will be a crash of the U.S. stock market.

My forecast is that, starting July 2015, stock markets across the world would experience extreme volatility and, by October 2015, the U.S. economy would experience an event like the Black Monday of 1987. As I have explained, since the 1980s, the U.S. has not been solving its macroeconomic problems but has been delaying the fix to its monetary policy that would bring back free markets so that wages track with employee productivity. The real fixes to our economic problems are being delayed by excessive monetary printing in the name of quantitative easing (QE). Hence, the coming economic crash of 2015 will most likely be a complete collapse of crony capitalism in the U.S. No matter how much the Fed delays hiking benchmark interest rates, the macroeconomic cycles of nature cannot be controlled by a country's central bank and the central banks cannot prevent the coming stock market crash by the end of 2015.

The role of macroeconomics in business can be seen in the way the condition of the economy affects individual businesses. For instance, during a recession, the behavior of consumers of goods and services changes to reflect the changes in the economy. Such changes can be seen in the way the demand for goods and services drops and the manner in which such a reduction affects the balance sheets of the various businesses. It is clear that the existing fabless-foundry business model of the U.S. Semiconductor industry does not comply with macroeconomic cycles of nature because it has not been able to minimize the problem of unemployment in economic downturns. What then should the business model be for the semiconductor industry to ensure that it complies with these macroeconomic cycles of nature, which are beyond the control of any human being, corporation, central bank, or nation? We shall explore this in the next chapter of this book.

4.5   REFERENCES

[1] Mulay, Apek, *Mass Capitalism: A Blueprint for Economic Revival,* Book Publishers Network, Bothell, WA, 2014.

[2] Mulay, Apek, "Transformation of U.S. Semiconductor Industry," Speakout, April 5, 2014. http://www.truth-out.org/speakout/item/22910-transformation-of-

`us-semiconductor-industry`

[3] Mulay, Apek, "The State of the Global Economy," May 13, 2015. `http://www.truth-out.org/speakout/item/30774-the-state-of-the-global-economy`

[4] Wikipedia, "Macroeconomics."

CHAPTER 5

Macroeconomic Cycles and Business Models for the Progress of Moore's Law

5.1 INTRODUCTION

As mentioned in the previous chapter, although the fabless-foundry business model is superior to the IDM business model, since commonsense macroeconomic parameters have been violated, the fabless-foundry business model has become a major contributor to U.S. trade and budget deficits. Hence, this fabless-foundry business model needs to be tweaked so that macroeconomic losses to the U.S. economy are rectified for a sustainable progress of Moore's Law. No progress is sustainable without a sustainable macroeconomic progress. Hence, the existing fabless-foundry business model needs to be tweaked so that not only does the semiconductor business model become sustainable, but it even complies with the macroeconomic cycles of nature that are beyond the control of any human being, corporation, or nation.

5.2 EXAMINING THE IMPORTANCE OF MACROECONOMIC CYCLES AND THE SEMI MARKET

The semiconductor industry has a potentially critical part to play in mitigating the effects of the natural macroeconomic cycles that occur in the United States economy. As explained in Chapter 4, these macroeconomic cycles, which last 7–14 years, are beyond the control of any human being, corporation, or nation. We need a business model for the semiconductor industry that takes into consideration these cycles, and avoids huge unemployment during economic downturns. Of course, high unemployment created during an economic downturn translates into substantial costs for the democratically elected government in providing unemployment benefits.

As the government works to minimize the spending during an economic downturn, the semiconductor industry must also work to minimize the problem of unemployment. Hence, the semiconductor industry business model should be able to establish a true free-market economy where the wages of employees automatically catch up with employee productivity in spite of economic downturns. Only when the wages catch up with productivity can supply and demand grow or fall in proportion. When this kind of a balanced economic growth is achieved, there will be a minimal overproduction of electronics in an economic downturn.

By minimizing the overproduction of electronics, layoffs could be avoided. As explained in Chapter 4 of my book *Mass Capitalism: A Blueprint for Economic Revival*, layoffs occur when real wages fail to catch up with productivity and lead to a poor consumption due to the overproduction of goods. Today, we need a true free-market business model for the U.S. semiconductor industry to ensure the broader prosperity and growth of the entire industry, and the existing defects present in the fabless-foundry business model, which result in trade and budget deficits, need to be rectified. This new business model should also ensure a robust free-market balanced economy that would boost the purchasing power of consumers.

When consumer purchasing power increases, then economic demand grows. This ensures that the consumers spend, which encourages the producers to make further investments to cater to the growing consumer demand. This kind of co-operative interaction between the producers and the consumers creates jobs. This is how the real job creators in a free-market economy are not only the producers but also the consumers. Since there is a growth of real wages in the economy, as compared to growth in demand from a growing consumer debt, resulting economic growth becomes very sustainable. This was precisely the miracle that occurred in the United States from 1950–1970 when there was a 4% year-over-year growth in the GDP—considered the golden era of free-market capitalism.

5.3   A THREE-TIER FABLESS-FOUNDRY BUSINESS MODEL FOR THE SEMICONDUCTOR INDUSTRY

It is an open secret that, for a variety of reasons, the U.S. manufacturing base has sharply deteriorated over the past three decades, and the semiconductor industry is no exception to this trend. In fact, this industry may have suffered more than some other American enterprises because of its capital intensive nature. The purpose of this section is to explain the causes of this decline and offer some commonsense economic policies that may lead to the industry's revival.

In the microelectronics area, a semiconductor fabrication plant (also called a fab) is a factory where such devices as integrated circuits are manufactured. A business that operates a semiconductor fab for the purpose of fabricating the designs of other companies, such as fabless semiconductor companies, is known as a foundry. If a foundry does not produce its own designs, it is known as a pure-play foundry. As of today, the semiconductor industry follows multi-national corporations' (MNCs) business model based on globalization. Following WWII, the United States pursued globalization, believing that American firms would be able to capture foreign markets. But the opposite happened. Other nations imported technology from gullible American companies and, with their low real wages, out-competed U.S. firms all over the world. The rest is history. By now, many American industries have disappeared, while most others have shrunk, resulting in a loss of jobs and stagnant wages.

When the Great Recession struck in 2007, the lingering weakness of the American economy, so far ignored by experts and the government, came to the surface. The present state of the U.S. economy has been very discouraging to the millions of Americans who are unemployed.

This jobs crisis is a result of the diminished purchasing capacity of consumers due to U.S. trade and macro-economic policies. With decreased exports and increased imports and the high cost of maintaining U.S. war fronts abroad, the budget deficit has been increasing over the years, causing serious concerns and political tensions. Additionally, counterfeit electronics entering the U.S. supply chain from abroad have raised national security issues.

Since the 2007 recession, the semiconductor industry has observed a relatively flat growth in its revenue and many small businesses have experienced a slowdown. Such stagnation has been analyzed in my recent book *Mass Capitalism: A Blueprint for Economic Revival*. In this chapter, I offer new ways to transform the entire semiconductor industry, enabling it to have high profits and steady growth. This model would help position the semiconductor manufacturing sector to break away from its usual boom-and-bust cycles and generate high profits in years to come.

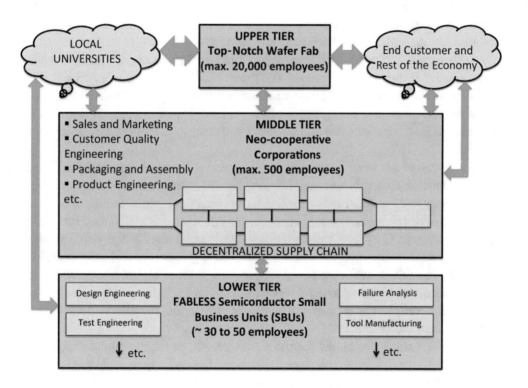

Figure 5.1: Three-tier fabless-foundry business model for the semiconductor industry that would lead to a balanced economic growth and increase consumer purchasing power. The upper tier has a top-notch wafer fab, which needs huge capital investments, the middle tier has neo-cooperative corporations and the lower tier has fabless semiconductor small businesses.

The proposed three-tier business model is designed to: (1) help restore a balanced economy without having to rely on overseas investment or foreign debt; (2) establish a free-market economy where the supply and demand of goods rise and fall automatically with minimal government intervention; (3) solve the problem of unemployment and hence excessive government spending that leads to budget deficits; (4) help the semiconductor business be at the leading edge of technology through sustainable capital investments; and (5) ensure a competitive business environment that would stimulate the growth of small businesses, and so on.

A balanced economy is critical to take the global semiconductor industry to its next level of innovation and financial success. An economy is considered to be balanced when there are no trade deficits and no budget deficits. Such an economy would increase domestic consumer purchasing power by letting the wages of workers catch up with their productivity, thus establishing an economic democracy. An economy is considered to be balanced when the supply of the goods from producers and demand from consumers grow and fall in the same proportion. Such an economy does not lead to overproduction, nor does it lead to any underutilization of available resources. The wages of the consumers contribute to demand and the productivity of the workers contributes to supply. A balanced economy ensures that wages catch up with productivity, which results in demand catching up with supply. As employees in a balanced economy work hard by being productive at their jobs, their wages also grows and these employees as consumers generate greater demand. This enables producers to manufacture more goods to meet the ever-growing needs of the consumers due to higher consumer purchasing power. Here there are no economic downturns, because consumer demand is always sufficient to match producers' supply of goods and services.

In addition to a balanced economy, we also need a decentralized supply chain, which increases co-operation among businesses, decreases wealth concentration in the economy, and improves efficiency and customer satisfaction. Macroeconomic reforms leading to wholesome decentralization of fabless business would boost the growth of several small fabless firms in the semiconductor ecosystem. Only with a decentralized supply chain does an individual player flourish. Hence, the decentralized supply chain leads to smaller organizations and a lower probability for mergers and acquisitions. A decentralized supply chain leads to higher co-operation among entities in an industry. By placing a higher value on building a relationship with the end customer, decentralization leads not only to overall customer satisfaction, but also to healthier long-term relationships with end customers.

I propose a three-tier model for the robust growth of fabless semiconductors based on economic democracy with economic decentralization. The entire semiconductor industry should be split into economic sub-systems. These sub-systems should be based on the availability of the necessary raw materials. This model is based on the theory of *Mass Capitalism*.

5.4 UPPER TIER: SEMICONDUCTOR MANUFACTURING FROM TOP-NOTCH WAFER FAB

A top-notch wafer fab needs an investment larger than a nuclear reactor. In 2013, the cost of building the next generation wafer fab was estimated at over $10 billion. Strategicly, the location of a wafer fab should strategiccally have ready availability of all necessary raw materials needed for manufacturing. The local government should build necessary infrastructure, like a domestic or international airport, good transportation facilities, good infrastructure, etc. for a smooth delivery of the goods to end customers. This would ensure the growth of small- and medium-size businesses, which would cater to that fab. Any infrastructure investment would also be a long-term investment to attract other businesses to that location.

In order to have a balanced economy, the official monetary policy should be such that wages keep up with labor productivity. Since workers' wages contribute to consumer demand and workers' productivity contributes to the supply of goods, when wages catch up with productivity, supply and demand grow and fall in proportion. Hence, it is important that company profits are first shared among the employees in proportion to their productive contributions, and later with any outside investors. In order to ensure that wages catch up with productivity, there should be special incentives offered to highly productive employees. This portion of profits should be allocated to those employees who are much more innovative and productive than others. The remaining profits, if any, should be shared with the private investors as a return on their investments. It is very important for an economy to first ensure that wages catch up with productivity to maintain a rational distribution of wages. This would also eliminate any economic imbalance that might occur as a result of huge income inequalities, which may result in economic recessions and depressions.

While company profits should be shared across the board in proportion to productive contributions of employees, these fabs, although government backed, should have complete autonomy to lay off any lazy or un-productive employees. This model would ensure that the wafer fab remains a top-notch fab at the forefront of innovation. Additionally, it would rectify the inefficiencies that may have existed in government-run businesses. Similarly, by letting the employees' wages catch up with their productivity, it would ensure a high consumer purchasing power in the economy.

For the semiconductor industry to be financially successful, it is critical that money circulates in the economy and does not remain idle in bank vaults or other forms of valueless hoarding. To make this feasible, it would be best for a wafer fab to offer retirement plans to its employees such that employees can invest some of their income towards the growth of their company by purchasing company shares. In this manner a wafer fab can raise much-needed capital, while its employees become part-owners.

This model has two benefits. Since employees own some shares of the wafer fab, they would work hard towards the success of the foundry. In economic downturns, these wafer fabs would prefer to share losses by taking across-the-board wage cuts or by cutting work hours of workers rather than laying them off. This would essentially maintain a balance between the supply and demand of goods. Additionally, since massive job losses due to layoffs could be minimized in economic downturns, it would also minimize government spending and budget deficits caused by offering unemployment benefits to laid off workers.

In order to engender accountability and transparency in employees' performance reviews, individual departments inside the wafer fabs should be decentralized so that different engineering teams are grouped in smaller teams that are able to elect their respective representatives to the corporate management board in a democratic way. This would enable employees to offer their innovative ideas as well as voice their concerns directly to the management through their representatives. These employee representatives would work with other members of the management board to ensure that the majority of decisions are taken in the best interest of employees, and hence of the company, thereby engendering a healthy relationship between the employees and management.

A good wafer fab should also collaborate with local universities by offering co-operative internships to engineering students and technicians. This model would enable a foundry to offer challenging thesis projects to students pursuing masters degrees in engineering through co-operative internships. The wafer fab should also offer doctoral research fellowships or internships so that these doctoral students would essentially work for the company at a fraction of the salary of regular research employees; this is the most economical way to fund R&D.

Local universities and foundries could also collaborate to arrange a mechanism by which:

1. Doctoral students' research could be sponsored by this "university-companies conglomerate," and then the resulting technologies (developed by the doctoral students) are spun off into new companies which are partly owned by this conglomerate and partly by the doctoral students and their supervisors. They would become future entrepreneurs of the semiconductor industry and be a part of the lower tier of this business model. Such conglomerates open up equal opportunities for the material and intellectual advancement to all students.

2. The "doctoral students owners" are then free to employ their master's degree graduates in their companies and implement a system of employee profit-sharing. This would not just help students with financial aid, but also help the semiconductor foundry get its R&D work done by the doctoral students.

5.5   LOWER TIER: THE FABLESS SEMICONDUCTOR BUSINESSES

An established top-notch wafer fab would also create local businesses, which provide necessary tools, test equipment, engineering services, etc. for the smooth functioning of the wafer fab. These providers of different services, in addition to the fabless semiconductor businesses, would form the lower tier of this three-tier business model.

In order to ensure sufficient job creation in the local economy, the fabless industry should undergo wholesome decentralization when it comes to offering these engineering services. Small business units (SBUs) should offer engineering services such as circuit design engineering, circuit layout engineering, test development engineering, failure analysis, tool manufacturing, and maintenance, etc.

Each engineering service provider should work as an independent SBU with a maximum of 30–50 employees in each business unit. These SBUs should comply with anti-trust laws, which should be strictly enforced to avoid mergers and acquisitions (M&A). Some of the existing fabless corporations are too big and they use their hefty profits to acquire small businesses. These M&As restrict competition, which has resulted in monopoly capitalism instead of the free-market-capitalism in the semiconductor industry. To avoid this opportunistic behavior the existing shares of all major corporations should be given to the employees of these corporations in proportion to their productive contributions. The next step should be to decentralize the operation of fabless companies and make each individual business unit function independently. This means the design engineering team would become an independent business; so would the product engineering team, customer quality engineering, reliability engineering, and so on. The respective businesses would operate at either lower tier or middle tier depending on their operation as shown in Figure 5.1.

The integrated device manufacturers (IDMs), which are currently privately owned and are not able to manufacture the latest technology products because of huge capital needs, should also consider splitting their fab and fabless businesses in accordance with the previously mentioned approach. The older fabs could function as tier 1 wafer fabs for analog chips, which do not need cutting edge transistor technology. The fabless businesses in these IDMs could be split into independent businesses to join either the middle or the lower tier, based on their business type. Similarly, large tool manufacturing corporations should be broken into SBUs to enable them to operate as independent small businesses. The wafer fab should give equal opportunities to all SBUs to compete for business to usher in a competitive free-market economy. This would encourage new entrepreneurs to start their own businesses to provide various engineering services to the principal foundry at tier 1, and hence promote innovation. The middle tier of the three-tier model should act as a medium to offer these services to the wafer fab and should also act as the most important sector, linking the end customer (or user of electronic products and services) to upper and lower tiers of the semiconductor industry.

All engineering departments involved at the pre-silicon and post-silicon stages should have a healthy competition with one another. This would enable the end customer to get products manufactured at significantly lower costs. Such a decentralization of the fabless business would provide the most innovative designs of new products and also eliminate any chance of large firms exercising their monopoly power.

Although the quality of products within some of the economic sub-systems would not be as great as those of other economic sub-systems, by having bench marking and knowledge sharing through international symposiums, publications, and discussion forums like Semicon, EBN, etc., it would be possible to gradually improve the quality of products and services for the entire global semiconductor industry in all subsystems. This is the only way that the global semiconductor industry could prosper along with the growth of regional semiconductor industry. In order to boost the development of more fabless SBUs for new entrepreneurs, the local government should help set up SBU associations all over the economy that could lease equipment and tools for setting up small businesses like failure analysis labs that require significant capital investment.

5.6   MIDDLE TIER: THE EMPLOYEE-SPONSORED NEO-COOPERATIVE CORPORATE SECTOR

The middle tier should include those semiconductor businesses that work directly with end customers. This sector should include relatively mid-size corporations with a maximum of 500 employees. This sector would interface directly with end customers, the upper business tier and the lower business tier. It would consist of neo-cooperatively managed semiconductor companies, where at least 51% of company assets are owned by company employees. All corporations in this tier should have exchange relationships as a decentralized supply chain. In a decentralized supply chain, individual units make decisions based on local information. In such a system, it becomes easy to incentivize players to act in co-operation, making the entire supply chain efficient.

The sales and marketing division would be able to get feedback on the demands from local customers and draft customer specifications to manufacture customized electronic gadgets based on the needs of the domestic economy. This middle-tier team would also interface with lower-tier SBUs, e.g., design engineering in order to develop customized electronic products as shown in Figure 5.1.

The packaging and assembly of chips would also be done in this tier. This tier would work on voluntary cooperation amongst corporations in the middle tier. However, those corporations that follow the model of employee-sponsored corporations should be given tax incentives in order to attract the other players in the middle tier to follow the model of employee-owned corporations. Majority shares of mid-size corporations in the middle tier would be owned by employees for them to have a stake in the success of their business. Since the middle industrial tier interfaces with both upper and lower tiers of the economic sub-system, it would be managing supply and demand of consumer electronics to maintain a balanced economy.

There are many advantages to have the middle tier in the semiconductor industry. If neo-cooperative corporations in this sector notice that customer demand is falling, then they would be able to communicate with the wafer fab at the upper tier and the fabless business unit at the lower tier to avoid overproduction of silicon. Both the upper and lower tier could utilize economic downturns to cut work hours of their employees, retrain their employees, or to concentrate on R&D activities. In the present global economy, due to the absence of the middle industrial tier, whenever there is a stock-piling of inventories in a wafer fab, the foundry lays off its employees because of poor economic demand. In this way, the middle tier can minimize job losses caused by macroeconomic cycles of nature.

With this middle tier, the neo-cooperative corporations would be able to adjust the supply and demand of electronics by means of cooperative actions of producers and consumers. Additionally, the middle industrial tier would provide an accurate real-time estimate of consumer demand, thereby providing feedback to upper and lower business tiers about customer requirements and demands in order to provide better customer service.

Another advantage of this three-tier business model would be "parallel processing," which is very much needed as the industry is progressing to adopt advanced transistor technology nodes. The middle-level corporate sector would be able to negotiate a good price for pre-silicon and post-silicon services and get the work started simultaneously with shorter lifecycles. This way manufacturing cycle time could be reduced and manufacturing costs would also decline significantly. Since the majority of corporate shares of this sector are owned by employees, there would be shared growth and prosperity, which would minimize the concentration of wealth in the hands of the few.

When the economy grows, the wages of all employees would also grow, leading to the growth of overall economy. The growth of overall economy would also grow the stock values of the corporations, and hence the year-over-year profits of all the employees in these corporations (not just CEOs, CFOs, and board of directors) would also grow steadily. When the economy slows down, the corporations would reduce work hours across the board to avoid job cuts due to layoffs. This way the problem of unemployment would be permanently solved. Additionally, since the middle tier is an employee-sponsored corporate sector, all majority corporate decisions would be made democratically by company employees, without any kind influence from non-employees.

A decentralized supply chain in the middle tier would generate high growth and employment without large-scale migration from rural to urban centers. This would avoid urban congestion and myriad related problems. Such a supply chain also engenders better customer satisfaction by guaranteeing product delivery through an alternate route in cases where the regular supply chain is disrupted by unforeseen events like natural disasters and social or political instability.

This three-tier business model for the global semiconductor industry would wholeheartedly accept automation in the industrial sector. Due to the use of new machines, productivity would grow exponentially because the supply of goods into the economy would also grow. In order to maintain the economic balance, consumer demand would have to match the growth in supply. In such a scenario, the neo-cooperative corporate sector would be able to meet the required production target with fewer work hours but pay its workforce a higher salary in proportion to their higher productivity resulting from the use of machines. This would give sufficient time for employees to pursue further education and vocational training, and help the workforce stay up to date with the desired skills needed to continue their careers in the ever progressing and rapidly advancing semiconductor industry.

This business model would make contributions to completely automate production of semiconductor chips from the beginning to the end, which is often referred to as "lights-out-fab." Such a business model would not only lead the fabless semiconductor industry to its next level of innovation and financial success, but would also act as a model for other sectors in the economy, leading to a vibrant growth of both regional and national economies.

5.7   CONCLUSION

When the semiconductor industry transitions to the three-tier fabless-foundry business model for sustainable progress of Moore's Law on the economic front, this model would also eliminate any problems of unemployment created due to macroeconomic cycles. The absence of a middle industrial tier in the present economy means that there is no premonition of change happening at a macroeconomic level, resulting in overproduction of electronics followed by layoffs due to poor consumer demand.

So far, the U.S. semiconductor industry has not been able to solve the problem of unemployment. Meanwhile, a microelectronics revolution has achieved two–four times the relative impact on the U.S. economy compared to the impact of the steam engine revolution almost 250 years ago. If the global semiconductor industry does not transition to a three-tier fabless foundry business model, the profitability from the progress of Moore's Law would get stalled due to poor consumer demand and, based on the macroeconomic cycles, the U.S. economy would take nearly 14 long years to come out of its economic depression, due to a growing gap between wages and productivity.

5.8   REFERENCES

[1] Mulay, Apek, *Mass Capitalism: A Blueprint for Economic Revival,* Book Publishers Network, Bothell, WA, 2014.

[2] Mulay, Apek, "The Causes of Economic Depressions," LinkedIn, May 6, 2014. https://www.linkedin.com/pulse/20140506053708--11893233-the-causes-of-economic-depressions

CHAPTER 6

Would Economics End Moore's Law?

6.1 INTRODUCTION

In 1965, Intel co-founder Gordon Moore, in "Cramming more components onto integrated circuits" in *Electronics Magazine* (April 19, 1965), made the observation that, in the history of computing hardware, the number of transistors on integrated circuits doubles approximately every eighteen months. This law is now used in the semiconductor industry to guide long-term planning and to set targets for research and development. This means that the cost of computer memory and computing power declines by 50% about every eighteen months. With the progression of transistor technology to meet Moore's Law, consumers today get more powerful tablets, smartphones, and other electronic gadgets. This empirical growth of components on an integrated circuit is known as Moore's Law. An explanation of the benefits from the progress of Moore's Law can be understood from the powerful consumer electronics gadgets in the hands of human beings. It is estimated that today's cell phones have more computing power than what NASA used to go to the moon in the 1960s.

6.2 DARPA'S BOB COLWELL ABOUT THE END OF MOORE'S LAW

<iframe src="https://www.youtube.com/watch?v=JpgV6rCn5-g" width="420" height="315" frameborder="0" allowfullscreen="allowfullscreen"></iframe>

Figure 6.1: Bob Colwell's Keynote speech at the Hot Chips 25 Conference entitled "The Chip Design Game at the End of Moore's Law," where Bob makes a compelling statement about the end of Moore's Law due to Economics before Physics.

On August 26, 2013, at his keynote speech at the Hot Chips 25 Conference entitled "The Chip Design Game at the End of Moore's Law," Bob Colwell, Director of the Microsystems Technology Office (MTO) at the Defense Advanced Research Projects Agency (DARPA), made a compelling statement about the end of Moore's Law: *"When Moore's Law ends, it will be economics that stops it, not physics. Follow the money."* Colwell also provided an indirect insight into reasons why Intel did not get into the mobile business early. He mentioned that the ones who are high-level decision-makers for those huge financial decisions at Intel Inc. should also take into consideration the architectural benefits of new designs. Unfortunately, as per him, Intel's culture does not support it, and hence Intel was late in getting into the mobile business.

Since the U.S. semiconductor industry violated commonsense macroeconomics, along with reforms in macroeconomy, the business models of the semiconductor industry also need to undergo reforms in order to surpass the economic limits of Moore's Law. In the case of Intel Inc., the company would not have been late entering the mobile market if its management had been more encouraging in getting input from its engineers. Engineers understand technology much better than most boards of directors, who primarily make financial decisions for a company. If the business model of an industry is reformed such that employees have a say in the operation of a business, today Intel Inc. would have been a dominant competitor of Qualcomm Inc. and other fabless players in the smartphone market. In that case, Intel's engineers would have been able to influence management decisions on upcoming technology and needed financial investments. But instead, today we are noticing layoffs at Intel Inc. and idle time increase for its fabs, which is a big loss for the capital-intensive wafer fab due to an accelerated depreciation of its state-of-the-art tools.

Bob Colwell also brought to the notice of conference attendees a very important life blood of business—the ability of a semiconductor company to forecast economic demand—but he did not offer any insight as to why companies fail at this. Commonsense macroeconomics teaches us that the ability to forecast demand for products needs the establishment of a true free-market economy. Only in a free-market enterprise system does the economic demand for various goods and services keep rising with constantly increasing consumer purchasing power. Thus, the consumer's purchasing power is a true engine of economic growth and a most reliable metric of economic performance as compared to conventional metrics, such as the GDP.

6.3   HOW CAN THE U.S. FEDERAL RESERVE HELP DOMESTIC ELECTRONICS MANUFACTURERS REDUCE U.S. TRADE DEFICITS?

According to a preliminary estimate of the Department of Commerce, US GDP grew at a mere 0.2% annual rate in the first quarter of 2015. However, economists now expect that this estimated growth was actually negative, thanks to a sharp rise in America's trade deficit in March 2015— from an estimated $35 billion to an actual $51 billion. This is exactly what I had feared would happen in my book *Mass Capitalism: A Blueprint for Economic Revival*. U.S. trade deficits have been slowly destroying American industry, including semiconductors and other high-tech products, ever since 1981. America's policy of free trade usually gets the blame for this deficit. That is why I had recommended that the United States should have balanced trade with its trading partners.

In a new book, *End Unemployment Now: How to Eliminate Poverty, Debt, and Joblessness Despite Congress*, one of the top American economists, Professor Ravi Batra, argues that the trade deficit actually results from two forces; while the United States follows free trade, China and Japan do not. These nations constantly intervene in the market for foreign exchange and manipulate their exchange rates in order to cheapen their currencies relative to the dollar. A cheap currency means cheaper prices for its goods abroad. In his recent volume *End Unemployment Now*, Professor Batra cites "mass capitalism as the wave of the future," where I have presented systemic problems caused at the macroeconomic level for the U.S. semiconductor industry as a result of America's free trade policies.

As explained by Professor Batra in this simple equation,

U.S. Imports from China =
China's Imports from the U.S. + China's Purchase of U.S. Government Bonds

If U.S. imports from China are $100 and China's imports from U.S. are $30, but China spends $70 to buy U.S. government bonds, then of course China's ownership of U.S. debt will increase. The growing ownership of U.S. debt by China has caused a lot of geo-political tension. The growing wage-productivity gap in the U.S., caused by its free-trade policies, has resulted in a waning manufacturing sector in the U.S. China also has a high wage-productivity gap, but they have avoided any over-production of goods by having an exchange rate with the United States that creates an artificial demand for dollars. In this way, China has not followed true free trade, and hence has avoided a depreciation of the dollar in spite of rising U.S. trade deficits. But in this process, China has become an increasingly larger owner of U.S. national debt.

For the U.S. to eliminate its trade deficits and for manufacturing to make a comeback in the United States, the U.S. needs to follow in the footsteps of China. It needs to manage its foreign exchange rate with the help of the Federal reserve so that the Fed offers an incentive to Chinese importers to buy more American goods. Professor Batra explains in his book that this can be made possible by offering a better exchange rate, like 4 Yuan for 1 USD, or perhaps even 3 Yuan for 1 USD. Let us take the example of an iPhone 6 manufactured in the U.S. Suppose, the cost of an iPhone 6 is approximately $500. With the current exchange rate of 6 Yuan for 1 USD, the cost of an iPhone 6 to a Chinese importer will be 3,000 Yuan. Now, suppose the Fed offers an exchange rate for China of 4 Yuan for 1 USD instead of 6 Yuan for 1 USD. Then the cost of an iPhone 6 to a Chinese importer would fall to 2,000 Yuan from 3,000 Yuan, which is a staggering 33% drop in price for the importer. If this does not work, the Fed can offer an exchange rate of 3 Yuan for 1 USD to increase US exports to China.

The Chinese importers would be delighted and there would be a rush to take advantage of the Fed's offer. China will buy more consumer electronics from the U.S. The U.S. would no longer run trade deficits but would gradually increase its exports to China. Eventually, American exports would soar to match imports from China. This is how more manufacturing jobs can be created within the United States and trade deficits can be eliminated with existing free-trade policies.

The free-trade policies do not allow the imposition of trade barriers, but do allow exchange rate manipulation. Professor Batra argues that when China and Japan have pursued such policies, why can't the U.S. pursue similar policies to reduce its domestic poverty, debt, and joblessness?

These trade policies would rekindle America's waning manufacturing base and create millions of high paying manufacturing jobs in the United States. The trade deficits would be eliminated without any major geo-political tensions. There would be no import duty placed on Chinese and Japanese electronics entering the United States. However, with the growing consumer purchasing power in the United States, the quality of Chinese goods entering the U.S. would have to automatically increase. The problem of counterfeit electronics would be eliminated when domestic consumers have sufficient buying power to demand good quality electronics. China's government should not resist this policy either, because Chinese exports to the United States would not be hurt, as a U.S. importer would still buy 6 Yuan for 1 USD from the People's Bank of China, and obtain Chinese goods at the same cost as before.

As long as all countries play by the rules, the process of offshoring manufacturing jobs just because of low costs in third-world countries would also come to an end. These policies could be adopted by countries like India, which run a trade deficit with China and hence have a huge problem of unemployment. Besides, China has also decided to compete with India's "Make in India" initiative to ensure that China retains its dominance as a global manufacturing hub. The Fed action would then create a free-market outcome, even though in reality there still would be no actual free trade, which would make it mandatory for countries like China and Japan to abstain from intervening in the markets. Once a balanced trade has been created between nations of the world, the economies can transition to Mass-Capitalism-based free-market economic reforms to further boost domestic purchasing power and hence increase domestic prosperity.

6.4   PHYSICS, NOT ECONOMICS, WILL END THE MOORE'S LAW

If we move beyond Moore's Law, the progress of other industries that depend on the semiconductor industry will also be slowed. The interrelation of the chip industry and this doubling of computer processing power has allowed for half a century of exponential growth. Bob Colwell sees 7 nm as being the end of the road (although not all experts agree) and predicts that Moore's Law will hit its limits around 2020 or 2022. It should be noted that ending Moore's Law at 7 nm happens because of its physical limits. However, the semiconductor industry is expecting to meet economic limits earlier than physical limits because of a transformation of the U.S. economy from free-market capitalism to monopoly capitalism. As opposed to free-market capitalism, where there is a healthy competition between businesses, in the case of monopoly capitalism, many industries (including the semiconductor industry) are dominated by a few giant firms that restrain competition.

As a result of monopoly capitalism, the consumer purchasing power of the majority in the economy has shrunk. There is a growing gap between the wages and the productivity of employ-

```
www.youtube.com/embed/CQoXXbVVZmg?list=UUelEJIyFJxfx9P2_9-4PXEQ
```

Figure 6.2: Book trailer video for *Mass Capitalism: A Blueprint for Economic Revival.*

ees in the global economy, which has resulted in a loss of economic balance. When capitalism is reformed to a free-market enterprise, and it works for all citizens in an economy, it results in an economic democracy. In order to sustain the progress of the semiconductor industry through the progress of Moore's Law, economic reforms become critical for semiconductor companies to justify their ever-increasing capital-intensive investments in the progress of Moore's Law. In my recent book *Mass Capitalism: A Blueprint for Economic Revival,* I put forth significant macroeconomic reforms along with financial, democratic, trade, and business reforms for the revival of the U.S. economy and particularly its semiconductor industry. Consider the following features of *Mass Capitalism* that would help address the present economic limits of the progress of Moore's Law:

1. *Mass capitalism* would ensure a fiscal and monetary policy in which there is no valueless hoarding of wealth by a few individuals and any valueless hoarding gets converted into valuable investments for sustaining the progress of Moore's Law.

2. It would ensure maximum utilization and rational distribution of all available resources in an economy.

3. It would optimize the business operations for the semiconductor industry in such a way that the potential of all employees would be properly utilized towards the progress of Moore's Law.

4. It would redesign corporate human resources policies in order to encourage optimum utilization of all employees' potential. However, organizations would also have to adjust properly to utilize that potential.

5. It would also ensure that the process of utilizing employees' potential is not the same for all employees of the semiconductor industry. While it would encourage better methods of utilization to be continually developed, the process of utilization would be progressive in nature.

If mass capitalism comes to reality, the result would be a robust growth of consumer purchasing power in an economy. By bringing back free markets, supply and demand would grow in proportion, thereby resulting in a balanced economic growth, low income taxes on individuals, higher investments, increased motivation for employees to work hard, and the growth of the overall economy. I believe that mass capitalism is the path forward for the U.S. and global semiconductor industry to reach its next level of innovation and financial success.

When these free-market reforms become a reality, even if the future improvements are less from one process generation to another, the macroeconomic growth in the overall economy would be very high. Through such profound macroeconomic reforms, the consumer purchasing power, and hence the prosperity of overall economy, would be very high. With higher economic demand, the demand for the latest and greatest electronic products would continue to grow. Such a robust consumer demand would force the semiconductor industry to make investments and manufacture products to meet that growing demand.

6.5   CONCLUSION

In this way, *Mass Capitalism: A Blueprint for Economic Revival* envisions sustaining the progress of Moore's Law in order to overcome its economic limits resulting from monopoly capitalism. These reforms would usher in the era of high prosperity and replace Colwell's hypothesis. Hence, If Moore's Law comes to an end, it will be due to physics, not economics.

6.6   REFERENCES

[1] Batra, Ravi, *End Unemployment Now: How to Eliminate Poverty, Debt and Joblessness Despite Congress,* Palgrave Macmillan Trade, New York, 2015.

[2] Colwell, Bob, "The Chips Design Game at the End of Moore's Law," Keynote Speech, *Hot Chips 25 Conference,* August 26, 2013.

[3] Mulay, Apek, *Mass Capitalism: A Blueprint for Economic Revival,* Book Publishers Network, Bothell, WA, 2014.

CHAPTER 7

Design of Supply Chains for the Success of the Internet of Things (IoT)

In today's globalized economy, supply chain professionals work together to deliver products to the end customer while working with different organizations and different business units within the same organization. During this process, supply chain managers face a challenge in controlling inventories and costs while maximizing customer-service performance. Before we understand the design of supply chains for success of the internet of things (IoT), let us understand the supply chains prevalent in the semiconductor industry and why the existing design of a supply chain will prevent the IoT revolution from taking off.

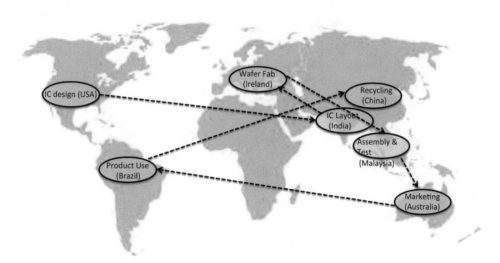

Figure 7.1: The supply chains of MNCs resulting from increased globalization of the semiconductor industry. Today, MNCs have spread their semiconductor production operations across the globe, including design engineering, test engineering, sales and marketing, packaging and assembly, etc.

A supply chain—a network of production and exchange relationships—spans multiple levels of production or task decomposition to let a producer buy inputs and sell outputs. Traditionally, supply chains have been formed and maintained over an extended period with extensive human interaction. Companies around the world base their business models on offering a range of products to satisfy the ever-changing consumer demand. A majority of the multi-national corporations (MNCs) have centralized supply chains. This brings the entire decision-making process to a centralized location (the MNC headquarters) and enables the MNC to centralize its production and profits.

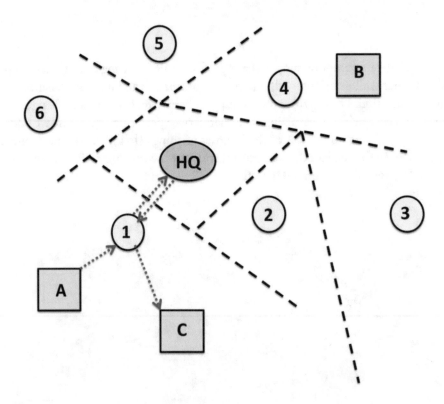

Figure 7.2: Centralized supply chain resulting in centralized decision-making and inefficiency. Here, the MNC has to transfer service from office **A** to office **C**. Both locations cater to the local head office **1** of the MNC, which is headquartered at **HQ**. In this centralized supply chain, to transfer service from office **A** to office **C**, the local head office **1** needs first to seek permission from **HQ** and then permit the service to be delivered from **A** to **C**. This makes the whole process highly inefficient.

In their paper "Material Management in Decentralized Supply Chains," published in 1993, Lee and Billington wrote "*Centralized control means that decisions on how much and when to produce are made centrally, based on material and demand status of the entire system.*" This approach turns out to be extremely inefficient for large corporations, leading to poor customer service. Another drawback is that a centralized supply chain leads to increased centralization of the entire economy, leading to high population densities in cities (more jobs at the centralized locations of MNCs) and rural economies going into the doldrums because of the chronic unemployment.

Decentralizing improve supply chain performance. In a decentralized supply chain, individual units make decisions based on local information. Such a system makes it easy to incentivize players cooperate, making the entire supply chain efficient. In today's globalized economy, MNCs are trying to capture market share through centralized supply chains. In these supply chains, all the profits from operations are transferred to a centralized location (corporate headquarters). This results in centralized planning and increases the concentration of wealth in the hands of a few.

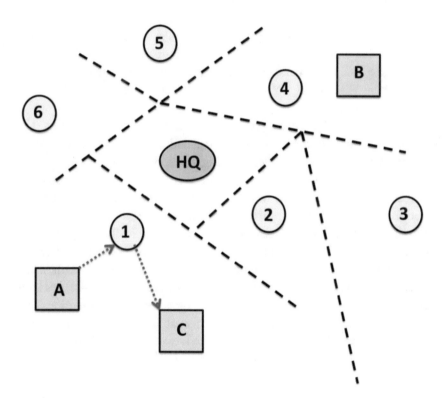

Figure 7.3: Decentralized supply chain results in more efficient decision-making based on local information. In a decentralized supply chain, to offer service from office **A** to office **C**, local head office **1** has complete autonomy and authority in making these local decisions, which makes the entire supply chain more efficient as compared with the centralized supply chain.

A decentralized supply chain has a lower wealth concentration, and offers a rational distribution of wealth that supports better macro-economic growth through higher consumer purchasing power and lower income taxes on local citizens. Additionally, since each individual player has a smaller influence on the entire supply chain, individual decision makers have less to gain by behaving in an opportunistic way.

Each individual decision maker has fewer profits to forgo by acting in an opportunistic way and has a broader customer base to gain by acting in the best interest of the supply chain. Hence, the decentralized supply chain leads to smaller organizations and lower probability for mergers and acquisitions. It also leads to higher cooperation over competition among different entities in the supply chain. By placing a higher value on building a relationship with the end customer, decentralization leads not only to overall customer satisfaction, but also results in healthier long-term relationships with end customers.

The decentralization of the supply chain also results in increased flexibility and decentralized decision-making. It engenders better customer satisfaction by guaranteeing product delivery through an alternate route in cases where the regular supply chain would be disrupted by unforeseen events like natural disasters and social and political instability. Decentralization has also demonstrated good performance in emerging market cooperatives, urban logistics, micro-retailing, etc. Decentralized design of supply chains would have an important role to play in the success of IoT revolution.

For the IoT to function effectively and create a vibrant local economy, the exchange relationships between the public and private broadband Internet providers, networking service providers, computational service providers, and big data service providers, as well as other software service providers, should be in form of decentralized supply chains. This would enable development of local internet protocol standards that cater to local economic needs. For example, to avoid communication overhead resulting from use of international standards like IPV6, the IP addresses could be simplified at local level, thereby making the communication faster and simpler between locally connected devices. This would also enable local companies to manufacture products and offer services that would cater to local economies, resulting in a vibrant growth of local economies. Decentralized supply chains would also ensure free-market economic reforms and thereby ensure "net neutrality" to function effectively with minimal government intervention. Decentralization of supply chains would also play an important role in enabling local economic planning for the establishment of robust broadband infrastructure. The local infrastructure should enable any device that is outside of local economy to be able to connect to the local network, through a switching interface (that handles the communication protocols at local level), thereby enabling a seamless user experience for any out-of-network user. In this way, decentralized supply chains would also ensure that the outflow of capital from the local area would be checked and, because it would remain in the local area, it would be utilized to increase production and enhance the prosperity of the local people. With the increasing demand for local commodities, large-scale, medium-scale, and small-scale IoT and other related businesses will all flourish.

7.1 CONCLUSION

The upcoming IoT revolution hopes to extend the end node far beyond the human-centric world to encompass specialized devices with human-accessible interfaces. As the IoT grows, the need for real-time scalability to handle dynamic traffic bursts also would increase. Without ensuring a prosperity of both producers and consumers, the IoT revolution can never succeed. Besides, creation of a vibrant local economy would boost the local economic demand and hence increase the supply of goods at a local level leading to a higher prosperity for all. Hence, decentralized design of supply chains offers a potential way forward for success of the IoT revolution.

7.2 REFERENCES

[1] Mulay, Apek, *Mass Capitalism : A Blueprint for Economic Revival,* Book Publishers Network. Bothell, WA, 2014.

CHAPTER 8

The Macroeconomics of 450 mm Wafers

SEMICON West 2014 in San Francisco, CA turned out to be a great place to meet bloggers in the semiconductor industry in order to get updated on the status of progress for 450 mm diameter silicon wafers. On one side, there is good news about the unprecedented level of collaboration taking place between the design and construction professionals through the Global 450 mm Consortium (G450C) to deconstruct a semiconductor facility matter associated with 450 mm adoption. But on the other hand, the transition to 450 mm seems to be delayed until 2020, with Intel and TSMC backing off their timing for the introduction of 450 mm wafers in volume production.

An analysis of the macroeconomics of this capital-intensive microelectronics business and its impact to macroeconomic growth of the U.S. economy shows that 450 mm diameter silicon wafers need to be introduced sooner rather than later to keep this business profitable through mass production and for sustaining its macroeconomic growth. This chapter presents an analysis of the macroeconomics of manufacturing 450 mm diameter wafers, how it would help in sustaining the progress of Moore's Law, and what precautions need to be adopted by the semiconductor industry at G450C to ensure a sustainability of huge capital investments for transitioning to 450 mm diameter wafers.

The global semiconductor industry has been constantly increasing the diameter of the silicon wafers it uses to reduce its manufacturing costs through mass production. The larger the diameter of wafers, the more real estate of silicon that is available for manufacturing. The increasing process complexities in nano-scale engineering add further to the silicon manufacturing costs. However, if the percentage increase in manufacturing costs per wafer from advancements in technology is smaller than the percentage increase in revenue from the larger real estate of silicon, only then does the semiconductor manufacturing business become profitable.

At present, the semiconductor industry is making wide use of 300 mm diameter wafers, and there is some progress towards 450 mm diameter wafers. But major players are delaying their investments in 450 mm diameter wafers due to significantly high manufacturing costs of semiconductor processing tools and lower than expected RoIs. The major players in this business do not expect to reap significant returns from their huge capital investments, which is raising questions whether mass production of 450 mm diameter wafers would become a reality anytime in the near future.

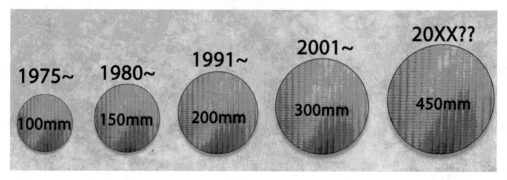

Figure 8.1: Timelines for the introduction of different diameters of silicon wafers. There is uncertainty regarding the introduction of 450 mm diameter wafers.

Whenever a new wafer size is first introduced, the cost per square inch of silicon on a given wafer size is at its peak. But, as the technology matures, this cost per square inch drops with time. The minimum silicon cost reached with 125 mm diameter wafers is about $1 (USD) per square inch. The minimum silicon cost with 200 mm diameter wafers is about $2 per square inch, resulting in a maximum cost per wafer of $100. The minimum silicon cost reached with 300 mm diameter wafers is about $3 per square inch, resulting in a maximum cost per wafer of $400. Thus, for every succeeding increase in wafer size, there is approximately a 1.4 times increase in costs of manufacturing.

The silicon wafers that are used for manufacture are sliced from a silicon ingot, which is approximately 99.99999% pure silicon. The increases in cost per square inch occur due to other reasons, such as the percentage of wafer that actually goes into the product, actual consumption of these wafers, and the R&D costs involved in the manufacture of wafers. Because of increasing processing complexities in subsequent generations of increased wafer size, the thickness of the wafers sliced from silicon ingots is larger. Thus, the total number of wafers that can be sliced from a silicon ingot decreases with increasing wafer sizes.

Additionally, the actual silicon going into the wafers is a small percentage of the total silicon available for use from these silicon ingots. According to Rose Associates, this was 30% for 150 mm wafers, 17% for 200 mm wafers, and close to 10% for 300 mm wafers. Based on this trend, the percentage of actual silicon from ingots that would go into use for 450 mm wafers would be even lower than 10%, possibly closer to 5%, based on the observed trend. Thus, the actual silicon from ingots going into electronic products is progressively decreasing for every generation of increased silicon wafer size. A larger wafer size also increases the number of chips that are mass produced.

The above data analysis educates us that, for ensuring a sustainability of huge capital investments in a transition to 450 mm diameter wafers, the majority of silicon that is manufactured inside a semiconductor wafer fab should get consumed in the manufacture of electronic products.

The question is, how does the semiconductor industry make it feasible? A good consumption of manufactured silicon can be achieved only when there is a robust economic demand for the latest and greatest electronic products. Hence, the G450C consortium should not only ensure a reduction in costs of manufacture to ensure sustainability of production, but should also collaborate to ensure that there is good consumption of its manufactured products. This can be made possible only in an economy where the consumers have a higher purchasing capacity to generate a robust demand for semiconductor products.

The G450C, a New York-based public/private program with leadership from Global-Foundries, IBM, Intel, Samsung, TSMC, and the College of Nanoscale Science and Engineering (CNSE), is housed on the State University of New York's (SUNY's) Albany campus and maintains focus on 450 mm process and equipment development. In this consortium, there is an unprecedented collaboration and cooperation between previous competitors in such a way that:

- all key players are coming together;

- CNSE is providing a uniquely neutral and technologically advanced home for critical research;

- work of the G450C is being guided by a strict application of an inside-out design approach; and

- key advances have been made in utility requirements, overhead conveyance systems, and energy efficient strategies.

This collaboration is a first step forward by the global semiconductor industry to ensure a steady supply of 450 mm diameter silicon wafers at a reduced cost. However, unless this consortium also ensures a similar cooperative collaboration to generate a steady demand for this newly manufactured silicon, the huge capital investments made by the participants in the G450C consortium towards a transition to 450 mm diameter silicon wafers cannot become sustainable.

On the macroeconomic side for the U.S. economy, the consumer purchasing power of the majority in the U.S. economy has shrunk because of the transformation of the U.S. economy from a free-market enterprise to monopoly capitalism. As a result of this, there is a growing gap between the wages and the productivity of employees in the U.S. and global economies, which is resulting in a loss of economic balance. When capitalism is reformed to a free-market enterprise, and it works for all citizens in an economy, it will usher in an economic democracy. As it would work for all citizens, monopoly capitalism will become mass capitalism.

In order to sustain the progress of the semiconductor industry that has been driven by the relentless progress of Moore's Law, macroeconomic reforms have become critical to allow semiconductor companies to justify their ever-increasing capital-intensive investments for transitioning to 450 mm diameter wafers. By establishing free-markets, supply and demand of electronic goods would grow in proportion, thereby resulting in a balanced economic growth, low income taxes on individuals due to smaller size of government, higher investments, increased motivation for employees to work hard, and the growth of the overall economy. These free-market reforms seem to be the only path forward for the global semiconductor industry to ensure its sustainability for the transition to 450 mm diameter silicon wafers.

When the G450C consortium collaborates to bring about such profound macroeconomic reforms, even if the future improvements in process technology for progress of Moore's Law are less from one process generation to another, the macroeconomic growth in the overall economy will be very high. These reforms would ensure that the consumer purchasing power and the prosperity of the overall economy would be very high. With a high economic demand, the demand for the latest and greatest electronic products will continue to grow. In this way, the G450C consortium can ensure a good consumption of all manufactured silicon from the larger size of 450 mm diameter wafers. This robust consumer demand would force the semiconductor industry to make more investments and manufacture the latest and greatest electronic products to meet that growing consumer demand.

8.1   COULD THE SEMICONDUCTOR INDUSTRY HELP THE U.S. COME OUT OF ITS THIRD ECONOMIC DEPRESSION WITH A TRANSITION TO 450 MM WAFERS?

There are two main causes for economic depressions—first, the concentration of wealth, and second, blockages in the flow of money. If capital is concentrated in the hands of a few individuals, most people will loose control over the economy and that control would be transferred into the hands of a wealthy few. This disparity results in the exploitation of a majority of the population, resulting in a serious depression. The concentration of wealth, is the fundamental cause of a depression.

Secondly, a depression may occur when money that is in the possession of individuals stops flowing. Money remains inert or unutilized because some wealthy folks who hoard a lot of wealth think that, if the money is allowed to flow freely, then their profits will decrease. In fact, their profits really increase when the consumer purchasing power of the common people rises due to the free flow of money. The very psychology of some wealthy individuals is to make a profit from avoiding free flow of money into the economy by hoarding a lot of wealth, instead of productively investing their wealth for economic growth. When wealthy individuals discover that the investment of money does not bring profits up to their expectations, then they stop investing. This keeps money immobile or inert; consequently, there is no investment, no production, no income, and hence no purchasing power. The situation becomes so dangerous that there are few buyers to buy commodities.

If there is surplus labor and deficit production, the effect of a depression is more acute. Hence, during a depression, the region having a surplus labor and deficit production would face an indiscriminate closure of business houses and lay-offs. When wages fall, the people in surplus labor areas, who used to go to deficit labor areas for employment, would face more hardships. This would aggravate the unemployment problem in surplus labor areas. Countries and regions with surplus production and deficit labor usually suffer fewer hardships during a depression.

Taking the above things into consideration, one can understand why the U.S. experienced the Great Depression of 1929 due to huge wage economic disparity. This also explains why it has becomes essential to establish a free-market enterprise system that would minimize the economic disparity with minimal government intervention.

Mass-capitalism-based free-market reforms envision the sustainability of huge capital investments for a transition to 450mm diameter silicon wafers. As explained above, it can be observed that an economic depression might occur when money that is in the possession of individuals stops flowing. Some of the major players in the global semiconductor business are concerned that their investments for transitioning to 450 mm diameter wafers would not give any significant returns. This could turn out to be one of the causes for money to remain inert or underutilized in this capital-intensive business.

As a result of the above policies, money in the economy would become immobile or inert; consequently, there would be no investment, no production, no income and hence further reduction in consumer purchasing power. The situation could become so dangerous that there would be very few buyers to buy new electronic goods. This macroeconomic analysis explains why macroeconomic reforms have become critical for transitioning to 450 mm diameter silicon wafers to ensure that money does not remain inert and keeps circulating in the economy so that consumer demand for the latest and greatest electronic products keeps rising. Without the above proposed macroeconomic reforms, progress of Moore's Law seems impossible and chances of the U.S. economy transitioning from its great recession to an economic depression seems inevitable. When mass-capitalism-based macroeconomic reforms enable a transition to 450 mm diameter wafers, then consumer purchasing power and consumer spending in the economy will grow. This would enable the United States to come out of its third economic depression, which started in late 2007, as explained in Chapter 4, "Impacts of Semiconductor Business Models on Sustainability."

8.2 REFERENCES

[1] Mulay, A., "Macroeconomics of 450 mm Wafers," Semi. org. July 31, 2014. `http://ww w.semi.org/en/node/50856`

[2] Mulay, Apek, *Mass Capitalism: A Blueprint for Economic Revival*, Book Publishers Network, Bothell, WA, 2014.

[3] Mulay, Apek, "The Causes of Economic Depressions," *LinkedIn*, May 6, 2014. `https://www.linkedin.com/pulse/20140506053708--11893233-the-causes-of-economic-depressions`

CHAPTER 9

Moore's Law Beyond 50

9.1   INTRODUCTION

In 1995, at his speech in the SPIE (The International Society for Optical Engineering) conference, Intel co-founder Gordon Moore hesitated to review the origins of Moore's Law and thereby restrict the definition of it. While resolving the contributing factors to the progress of Moore's Law, Moore realized that the semiconductor industry not only progressed to larger diameter silicon wafers, but simultaneously evolved into finer and finer dimensions. He attributed the contributions to increased complexity arising from increasing density of chips to be "*circuit and device cleverness.*" Moore attributed 60% of the progress of Moore's Law to die size and finer structures and the remaining 40% to circuit cleverness.

Although Moore acknowledged that it was hard to predict beyond the next couple of generations, he considers it to be a spectacular achievement of the semiconductor industry to stay on the exponential so long. However, Moore is concerned about the increasing costs of the equipment for every generation of technological progress. At his speech at the 1995 SPIE conference, Moore said:

> "*The people at this conference are going to have to come up with something new to keep us on the long-term trend. Capital costs are rising far faster than revenue in the industry. We can no longer make up for the increasing cost by improving yields and equipment utilization. Like the 'cleverness' term in device complexity disappeared when there was no more room to be clever, there is little room left in manufacturing efficiency.*"

This chapter explains the contributing factors to the progress of Moore's Law over the years and how this law could continue in this era of dwindling returns on investments.

9.2   POTENTIAL CONTRIBUTIONS OF THE SEMICONDUCTOR INDUSTRY TO THE GLOBAL ECONOMY

In 1986, the Semiconductor industry represented just 0.1% of gross world product (GWP). In 2005, the contribution of this industry accounted for 1% of GWP. The Industry stays on this path of progress and is able to sustain the progress of Moore's Law, the contribution to GWP would be close to 10%. The information technology (IT) revolution, with the IoT and big data being the major contributors to the next digital economy, has a potential to sustain the progress of Moore's Law, making it one of the biggest industries in this world.

Today, costs of manufacturing are rising exponentially and revenues earned are not able to grow at a commensurate rate. Given the importance of Moore's Law, it is very important to sustain its progress in order to sustain the progress of a knowledge-based U.S. economy. Gordon Moore summarized that the progress of Moore's Law could be attributed to the fantastic elastic market (a result of globalization) and technology that often exploits what Moore described as exceptions to Murphy's Law. Since the latter half of the 19th century, the U.S. has been able to achieve a regular non-inflationary growth of 4–5% due to information technology. This information technology is also energy saving.

Since Moore's Law resulted in higher productivity and reduced the costs of transistors due to mass production, the progress of Moore's Law has benefited the producers, due to its productivity gains. However, Moore's Law was also supposed to benefit the consumers, due to lower costs. As explained in the subsequent section, the progress of Moore's Law has benefited the producers much more than it has benefited the consumers. The value of Moore's Law has also been understated as a powerful deflationary force in the world's macroeconomy, when it comes to supply of goods into the economy.

On his retirement, Gordon Moore had the following to say about the future of the semi-conductor industry,

> *"I helped get the electronics revolution off on the right foot . . . I hope. I think the real benefits of what we have done are yet to come. I sure wish I could be here in a hundred years just to see how it all plays out."*

Moore's Law has continued to plug on, delivering benefits to many who will perhaps never appreciate the important contributions of this man and his observation. We shall explain in subsequent sections of this chapter why it is important for semiconductor industry professionals to be open minded to newer ideas that would let the benefits from the progress of Moore's Law reach a larger share of the global economy.

9.3   MONOPOLY CAPITALISM'S EFFECTS ON THE PROGRESS OF MOORE'S LAW

In 1968, if you were to purchase a 256 KB memory for a mainframe computer, it would have cost you $100,000. If you divide that total cost by 256, it comes to $391 for each thousand bytes of memory. Today, it costs $6 to buy eight gigabytes of memory, which comes to $0.00000075 for each thousand bytes. Simple math shows us that prices went down even faster than Moore predicted in his law, and you do not have to take into consideration rising inflation over the years. If costs of memory have gone down so much, Moore's Law's relentless progress over several decades should have contributed to reduced costs for energy, medicine, law, education, financial transactions, and governments. However, we have not seen these prices fall so much for all consumers. The plunging prices due to the progress in technology have mainly benefited the highest income earners because of monopoly capitalism. Due to monopoly capitalism, the corporations, through their lobbying efforts and being controlled by outside investors, have prevented plunging technology prices from benefiting all Americans.

9.4   A CASE STUDY OF MONOPOLY CAPITALISM IN THE U.S. TELECOMMUNICATIONS INDUSTRY

Here is an example of the telecommunications industry that demonstrates how the plunging technology prices have helped the producers more than the consumers. The amount of data that can be squeezed into copper, fiber, or radio waves at a given price in a given time decides the costs in the transfer of data. The cost of data-squeezing depends on the costs of computer power that does the squeezing. The dial-up Internet in the 1990s, which transmitted 32 kilobits of data per second, cost approximately $0.61 per thousand bits per second. Today, high-speed Internet enables 10 megabits per second of data transfer, which comes to $0.004 per thousand bits per second. However, with a one-time set-up cost in infrastructure, American consumers should see the price of Internet service fall every year with the progress of Moore's Law, which has not happened. Today, services like Skype enable users to video call around the world over the Internet for free. Users can transfer huge quantities of data on the Internet without worrying about how far it is going or whether it crosses the U.S. border. However, even though data communication with VoIP (Voice-over-Internet Protocol) needs the same amount of data to be transferred as compared to a comparable voice communication using a telephone network. The prices that consumers pay for data transfer using voice communication with the telephone network are much higher than the data transfer costs for voice and video communication through the Internet using Skype.

Advances in the speed of data transfer because of the relentless progress of Moore's Law over the years should also have decreased the cost of data communication. Instead, the high prices are due to government-owned or government-sanctioned monopolies in the voice telephony structure. The monopolies in the telecommunication industry lobbied the U.S. government and United Nations' International Telecommunication Union (ITU) to keep international calling rates high. The VoIP industry flourished, as it did not make use of any infrastructure from the monopolies in the old telephony system, and hence the Internet led to a revolution of an entire industry. There are still a few cell-phone service providers in the United States who do not let the radio-frequency usage prices fall because of their industry's monopoly. This is one example of how the progress of Moore's Law has not benefited consumers to the extent to which it should have because of monopoly capitalism.

9.5   OTHER INDUSTRIES TRYING TO IMITATE MOORE'S LAW

Monsanto's 1997 annual report proclaimed Monsanto's Law, which is "the ability to identify and use genetic information is doubling every 12 to 24 months. This exponential growth in biological knowledge is transforming agriculture, nutrition, and health care in the emerging life sciences industry." The measure of growth from Monsanto's Law can be found from the number of registered genetic base pairs, which grew from nil to almost 1.2 billion between 1982 and 1997. Magnetic memory has seen a similar parallel to Moore's Law as it shrinks the size of a magnetic pixel. Life science's gains are a direct result of increased modeling capability of increasingly powerful computers. Magnetic memory's gains are a direct result of chip manufacturing methodologies being applied to this field. Both are a direct result of the benefits gained from Moore's Law. Indeed, Paul Allen of Microsoft fame has credited his observation that there would be a need for more increasingly powerful software as a direct result of learning about Moore's Law, since the software rides on the hardware. He reasoned that this would be the outcome of increasingly powerful chips and computers, and then convinced Bill Gates there was a viable future in software—something no major systems maker ever believed until it was too late for them.

However, one thing is certain about the progress of Moore's Law so far—the progress of only the supply side of silicon—resulting in the formation of monopolies in not only hardware and software industries, but also in several other industries like genetic engineering, etc., because of ignorance of the common sense macroeconomics.

9.6   REDEFINING MOORE'S LAW AT 50

Since its inception, the semiconductor industry's ability to follow Moore's Law has been the engine of a virtuous cycle. Through transistor scaling, one obtains a better performance-to-cost ratio of products, which induces an exponential growth of the semiconductor market. This in turn allows further investments in semiconductor technologies, which fuel further scaling. The NTRS roadmap has transformed into the ITRS roadmap in the effort to scale transistor dimensions in proportion to Moore's Law for the continuation of this virtuous cycle. On one hand, this ITRS roadmap has helped to identify the knowledge gaps for this trend to focus the R&D efforts. But on the other hand, the increased globalization of the semiconductor industry has violated the macroeconomics parameters of the U.S. economy by means of increased offshoring of high-tech manufacturing jobs from the U.S. to low labor-cost countries (LLCs).

April 19th, 2015 was another important landmark for the global semiconductor industry because it was the 50th Anniversary of Moore's Law. Let us take a look at the definition of Moore's Law as defined by Gordon Moore on April 19th, 1965. It stated:

> *"The number of transistors per square inch on integrated circuits had doubled every year since their invention and this trend will continue into the foreseeable future."*

Moore's Law provided a predictable business model for the semiconductor business, where companies continue to keep investing on advancements in shrinking transistor dimensions resulting in increased productivity, profitability, and performance of ICs. A predictable business model for any industry ensures better RoIs for that industry. Hence, Moore's Law provided 50 years of predictable business modeling for the global semiconductor industry to contribute towards technological growth and may not be regarded as a self-fulfilling prophecy, as claimed by many semiconductor industry professionals. Over the last 40 years, Moore's Law has essentially held true, with the only difference being that the transistor doubling is now occurring every 24 months. Virtually every semiconductor company uses Moore's Law as a guide to develop their product strategy or roadmap to evaluate three important attributes:

1. Timing of introduction of new products;

2. Integration of new features to existing products; and

3. Cost of products.

Over last 50 years, the progress of Moore's Law has been scaling the transistor dimensions at all costs and ignoring macroeconomic parameters of the economy in this process. *No progress is sustainable without a sustainable macroeconomic progress.* Hence, for Moore's Law to continue benefiting today's knowledge-based economy, it would certainly need progress on the physical side, but would also need equally good progress on the economic side. Only when both physics and economics succeed can Moore's Law succeed after its 50th anniversary—until the physical dimensions cannot be scaled any further, thus bringing about a demise of Moore's Law due to physics, and not economics.

The revenue of the global semiconductor industry has grown at a very rapid rate and there has been hundredfold times growth of this industry from 1968–2004. Hence, on the 40th anniversary of Moore's Law, Gordon Moore said that his law is actually about economics. At that time, Moore further stated that:

> *"In fact, there was even a period during the 1970s when the industry was more than doubling the total number of transistors ever made every year, so that more electronics were built each year than existed at the beginning of the year. The pace has slowed recently but is still on a good growth curve."*

From its early beginnings in 1970s, the semiconductor industry followed the integrated device manufacturers (IDMs) business model. When the tech bubble burst in 2001, the increase in fab costs, combined with higher R&D costs due to the necessity of investing in both product and process development for the relentless progress of Moore's Law, resulted in the vast majority of semiconductor companies now adopting the "fabless-foundry" business model. This fabless-foundry business model has resulted in transforming the global semiconductor industry.

Under this fabless-foundry model, the semiconductor company focuses on product development, while the process R&D and the manufacturing capacity investments are done by third-party foundries. Each foundry enters into manufacturing supply agreements with several fabless players, which allows the foundry to scale up and obtain a return on its process R&D and capital investments. Additionally, it also enables small fabless players to concentrate on innovative designs and outsource manufacturing costs to foundries. This acts as a win-win for not only the fabless players but also for the foundries.

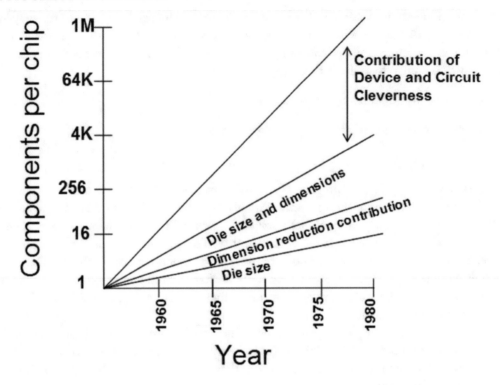

Figure 9.1: Resolution of complexity increase into contributing factors. *Source: Intel.*

In his publication, *Understanding Moore's Law: Four Decades of Innovation*, author David C. Brook presented all the contributing factors taken into consideration by Gordon Moore in the resolution of the complexity of integrated circuits. As shown in Figure 9.1, the plot of contributing factors to the resolution of complexity of integrated circuits, all the contributing factors have been only to the supply side of the silicon in the economy. However, there is no mention of growth in demand for this manufactured silicon to ensure its consumption.

The increased globalization of the U.S. semiconductor industry has transformed, the National Technology Roadmap for Semiconductors (NTRS) into the International Technology Roadmap for Semiconductors (ITRS). In this progress of Moore's Law, the Global semiconductor industry has ignored macroeconomic parameters and hence the consumer demand for silicon has started to slow down, due to reduced real consumer demand, which comes from a growth in the wages of consumers in proportion to their productivity. As progress cannot be sustained without a sustainable macroeconomic progress, the pace of growth has slowed recently, because of which semiconductor giants like Intel, TSMC, Samsung, etc. have stopped making investments in 450 mm diameter silicon wafers (which offers a potential path for generating more revenue for this industry). As long a semiconductor industry's progress complies with a sustainable macroeconomic progress, progress of Moore's Law beyond 50 years can continue to benefit the semiconductor industry and global economy. Today, Moore's Law needs to be redefined for sustainable macroeconomic progress for the progress of the law.

In my book *Mass Capitalism: A Blueprint for Economic Revival*, I have mentioned one of the five important features of this free-market theory of *Mass Capitalism* as:

> *"Time and space are changing, and the semiconductor industry will have to adjust with those changes. The principles of mass capitalism will not change; rather the application of these principles will have to adjust with the changing circumstances. The semiconductor industry will have to move forward by recognizing and adjusting with changes in time and space."*

Based on this feature and for sustainable macroeconomic progress, Moore's Law should also adjust to changes in time and space. Hence, for a predictable business model for the semiconductor industry beyond 50 years of Moore's Law, Moore's Law needs to be redefined. Moore's Law could be re-defined as:

> *"The macroeconomic policies should establish a true free-market economy in the U.S. semiconductor industry such that the continued growth in the number of transistors per square inch on integrated circuits grows in proportion to the consumer purchasing power in the economy, to ensure a sustainable consumption of the manufactured electronics and sustains the increased demand for more electronics. This shall ensure that doubling the number of transistors on integrated circuits would continue into the foreseeable future."*

9.7   REFERENCES

[1] Andrei A. Kirilenko and Andrew W. Lo, "Moore's Law versus Murphy's Law: Algorithmic trading and its discontents," *Journal of Economic Perspectives*, Volume 27, Number 2, pp. 51–72, Spring 2013.

[2] David C. Brock, *Understanding Moore's Law: Four Decades of Innovation*, Chemical Heritage Foundation. Philadelphia, PA, pp. 122, 2006.

[3] Hutcheson, Dan G., "Moore's Law: The History and Economics of an Observation that Changed the World," Electrochemical Society Interface, pp. 17–21, Spring 2005.

[4] Huff, Howard (Ed.), "Into the Nano Era : Moore's Law Beyond Planar Silicon CMOS," Springer Series in Material Science, Vol. 106, Series ISSN0933–033X, pp. 11–35, 2009.

[5] Moore Gordon E., "Lithography and the Future of Moore's Law," SPIE Speech, 1995.

[6] Mulay, Apek, *Mass Capitalism: A Blueprint for Economic Revival*, Book Publishers Network. Bothell, WA, 2014.

[7] Mulay, Apek, "Moore's Law Will Not Come to an End Anytime Soon," LinkedIn. March 15, 2014. https://www.linkedin.com/pulse/20140315035003--11893233-moore-s-law-will-not-come-to-an-end-anytime-soon?trk=mp-reader-card

[8] Mulay, Apekshit, "50 Years of Moore's Law - Chips Off the Old Block," *The Economic Times,* Apr 18, 2015. http://epaperbeta.timesofindia.com//Article.aspx?eid=31815&articlexml=50-YEARS-OF-MOORES-LAW-Chips-Off-the-18042015008037

CHAPTER 10

Macroeconomics of Semiconductor Manufacturing for Emerging Economies

10.1 INTRODUCTION

Having more developing economies participate in high tech manufacturing would not only help in creating more manufacturing jobs, but it would also help transform their economies into developed economies. A higher consumer purchasing power of individuals would help sustain the progress of Moore's Law when there is a competitive capitalism in the global semiconductor industry as compared to today's monopoly capitalism. In this chapter, we explore how semiconductor manufacturing facilities could become a reality in these developing economies. According to the International Monetary Fund (IMF), the Indian economy has a potential to surpass China as the world's leading economy in 2015, and there is a lot of optimism of the global investors because of India's thriving middle class. Besides, the Make in India campaign launched by India's newly elected Prime Minister, Narendra Modi, to promote native manufacturing will now focus on developing action plans for several key sectors, including automobiles and components, aviation, bio-tech, construction, defense production, electrical machinery, electronics, IT, media and entertainment, food processing, mining, oil and gas, ports, pharmaceutical, railways, and thermal power. Semiconductor manufacturing will play a key role in this Make in India initiative to reduce India's import bill on imported electronics.

10.2 "MAKE IN INDIA": TAKING VISION TO REALITY FOR INDIAN SEMICONDUCTOR MANUFACTURING

The government of India has approved setting up two semiconductor wafer fabrication (fab) manufacturing facilities in India. These fab units are to be set up by two business consortia, one led by Jaiprakash Associates Limited with IBM USA as technology partner [Technology: 90/65/45/28 nm, Capacity: 40,000 WSPM] and the other led by HSMC Technologies India Pvt. Ltd with ST Microelectronics as technology partner [Technology: 90/65/45/28/22 nm, Capacity: 40,000 WSPM].

The government of India will extend several incentives to the two consortia:

- A 25% subsidy on capital expenditure and tax reimbursement as admissible under the Modified Special Incentive Package Scheme (M-SIPS) Policy.

- Exemption of Basic Customs Duty (BCD) for non-covered capital items.

- A 200% deduction on expenditure on R&D as admissible under Section 35 (2AB) of the Income Tax (IT) Act.

- Investment-linked deductions under Section 35AD of the IT Act.

- Interest-free loan of approx. INR. 5124 crore each. (Exact amount to be calculated on appraisal of a detailed project report)

One way for India to reduce its current account deficits from its imported electronics is by attracting an inflow of foreign capital to India. This would compensate for any flight of foreign capital when the U.S. Federal Reserve starts hiking its benchmark interest rates. Since the financial crisis of 2008, countries like India have been prime beneficiaries of quantitative easing (QE) policies of the U.S. Federal Reserve. The investors borrowed cheap short-term money in the U.S. and invested the same in higher yielding assets in India. The massive capital inflows also let India comfortably finance its trade and current account deficits. However, now that QE has come to an end in the U.S., the capital inflows into India will start moving out, putting pressure on the Indian rupee in coming years. In India, the Reserve Bank of India (RBI) has been intervening in support of the struggling rupee in recent sessions, triggered by a worsening trade deficit.

As part of its Make in India initiative, India wants to become a global semiconductor manufacturing hub like China. In an attempt to build India's industrial base nationwide, Prime Minister Narendra Modi is pushing the Make in India campaign, designed to attract foreign investment by highlighting the ongoing changes. *"We have to increase manufacturing and ensure that the benefits reach the youth of our nation,"* Modi tweeted after the initiative's Sept. 25, 2014 introduction. With this vision in mind, Modi also invited U.S. President Barack Obama as a chief guest on the occasion of India's 66th Republic Day celebrations on January 26, 2015.

'Make in India' in high-tech semiconductor manufacturing is certainly an important step taken by India to minimize their current account deficits because of the large quantity of imported electronics due to its burgeoning middle class. However, without significant macroeconomic reforms, Make in India will not be able to solve its intended goal of reducing India's current account deficits. Let us analyze the problem before we look into proposed solutions for the success of Make in India.

First, from a macroeconomic perspective, there are two components to an account deficit viz. trade and budget deficits. India's free-trade policies have made a major contribution to India's trade deficit over the years. As a result of free trade, not only has India lost its major manufacturing industries to China but this has led to the entry of foreign manufactured goods into the country without having to pay any import duty, including imported consumer electronics. If India wishes to ensure that Make in India becomes a success for high-tech semiconductor manufacturing, it will have to replace its free-trade policies by means of fair-trade policies to ensure that domestic manufacturing can withstand foreign competition. If India is forced to pursue free trade by WTO then it will have to manage its foreign exchange rates with the help of RBI to increase its exports to countries like China, with whom India runs a trade deficit that results in high unemployment in India.

In addition, India should also make note of the fact that the U.S. Federal Reserve is on track to raise interest rates, which would also force the Fed to reform its monetary policies so that wages keep track with productivity to avoid an economic crash. The U.S. will not be able to run any trade deficits after the Fed starts to hike its interest rates. As a first step, the U.S. has already announced huge tariffs on solar goods imported from China and Taiwan. US-based MNCs will not be able to offshore manufacturing jobs to low labor-cost countries (LLCs) and import these goods into the United States without having to pay any import duty, in order to reduce U.S. trade deficits.

Hence, India should not depend on U.S.-based MNCs to offshore their manufacturing to India due to macroeconomic changes that are coming to the U.S. economy. Just as the U.S. has been backshoring manufacturing from LLCs to the U.S., to reduce its trade deficits, India should also plan on making Make in India successful by ensuring robust domestic demand for sustainable manufacturing. Therefore, even though the hourly labor costs in India for manufacturing averages $.92 as compared with $3.52 in China (according to Boston Consulting Group), the U.S. will not be able to sustain its trade deficits any longer by means of offshoring manufacturing to LLCs like India in order to retain the value of the U.S. dollar.

With regards to India's burgeoning budget deficits, these deficits are a result of monetary policies followed by India's central bank, which result in growing disparity in India. Just like the Fed, the Reserve Bank of India (RBI) also follows monetary policies such that real wages trail employee productivity. When the Indian government offers huge tax incentives to the two consortia for establishment of high tech semiconductor manufacturing facilities in India, and does not reform the monetary policies of its central bank, the net result of these policies will be an addition to India's burgeoning budget deficits.

The U.S. has been able to sustain its twin deficits by printing its currency, but the coming macroeconomic crisis in the U.S. would force the Fed to follow a monetary policy that ensures that wages keep pace with productivity. Unlike the U.S., India cannot sustain its deficits by printing currency, as India is not a global trade currency. Such policies have been responsible for rising inflation in the recent past. Once wages catch up with productivity in India, the resulting balanced economy will ensure minimal intervention by RBI when an increased domestic consumer demand calls for its own supply of goods.

If these commonsense macroeconomic reforms are not implemented by authorities involved in making the fabs a success for the Indian economy, the budget deficits will start to soar from huge tax breaks given to the consortia. Additionally, India's trade deficits will also soar from the import of duty-free consumer electronics due to its free trade policies. These twin deficits will be massive due to a growing middle class and would cause a disastrous failure of Make in India for high-tech semiconductor manufacturing. The present plan will not be able to achieve its goal of eliminating India's projected current account deficits of $400 billion (from imported electronics) by 2020. Only a balanced economy can help India make its huge investments sustainable so that the semiconductor manufacturing facilities will be able to cater to a robust domestic consumer demand.

Recently, the U.S. announced huge tariffs on solar goods from China and Taiwan as a first step to boosts its domestic manufacturing. Eventually, the U.S. will have to impose tariffs on all foreign manufactured goods entering the U.S. in order to boost domestic manufacturing and eliminate its trade deficits. In that regard, Indian manufactured goods will be not able to compete in the U.S. just because of their low cost of manufacturing when an import duty is imposed from fair-trade policies. Hence, India needs to ensure that it has a robust domestic demand for electronic goods so that the semiconductor wafer fabs remain in operation 24×7 to minimize any idle time.

The solution to this macroeconomic crisis is to reform the Reserve Bank of India (RBI)'s monetary policies to usher in a competitive free-market economy so that wages keep track with employee productivity and eliminate budget deficits. Additionally, by ensuring that wages keep track with productivity, a robust consumer demand can be ensured through the growth in real wages as compared to unsustainable demand through the growth in consumer debt.

The trade deficits have to be eliminated by reforming free-trade policies by fair-trade policies or by managing its foreign exchange rates with China in order to boost U.S. exports to China. In this way, these macroeconomic reforms, and by means of establishing a fabless-foundry semiconductor eco-system based on a three-tier business model, would lead to balanced economic growth and a sustainable semiconductor manufacturing ecosystem for India.

10.3   CAN "MAKE IN INDIA" BECOME SUSTAINABLE FOR THE INDIAN SEMICONDUCTOR MANUFACTURING SECTOR WITH COMING MACROECONOMIC CHANGES IN THE U.S. ECONOMY?

The semiconductor manufacturing is the most capital-intensive business and it is very important to make these investments sustainable in the short term in order to ensure profitability in the long term. Sustainability of the semiconductor wafer fabs involves being able to keep the fabs in operation 24×7 to reduce the tool idle time and to manufacture semiconductor wafers that meet the growing demand for consumer electronics and military needs.

My recent book *Mass Capitalism: A Blueprint for Economic Revival*, takes you on a journey of semiconductor manufacturing in the U.S. semiconductor industry. The high cost of manufacturing and keeping track with the ITRS roadmap to keep up with progress of Moore's Law has forced the offshoring of the IC packing industry, design engineering services, and, eventually, even the manufacture of semiconductor wafers from the United States to Asia. These policies of globalization have resulted in rising trade deficits for the U.S. The replacement of manufacturing sector with relatively low-paying service sector jobs has resulted in falling incomes and a depreciating middle class in the U.S.

In this way, globalization of semiconductor manufacturing resulted in a loss of dominance of the U.S. semiconductor industry and started to make this capital-intensive as well as knowledge-intensive business unsustainable, leading to an early demise of Moore's Law (due to economic limits because of huge capital investments) because of poor RoIs due to poor domestic consumer demand. Since the 2008 global financial crisis, the U.S. has been trying to revive its economy with its benchmark interest rates close to 0% and following QE policies to stimulate its economy. Instead of reviving the economy by boosting domestic consumer demand, the QE policies have resulted in growing income disparity, as the wages of the middle class haven't been growing to boost consumer demand.

The growth in domestic demand from increased consumer borrowing due to low interest rates is unsustainable as interest rates cannot remain low forever. Additionally, the low interest rates have not increased domestic investments in the U.S. and, instead, investors have preferred to get better returns on their investments by investing in developing countries with higher interest rates, like India. Hence, low benchmark interest rates in developed economies like the U.S. and Europe have primarily benefited the wealthy individuals in helping them get cheaper loans on mortgage properties and helping them earn higher incomes through renting these properties. These monetary policies haven't encouraged the easy money from QE to get invested in the U.S. economy, as investors have preferred to invest for higher yields in countries like India. Hence, QE policies have not been able to solve the problem of unemployment in the U.S. and have mostly created low-paying and part-time jobs in the U.S.

Now that the QE has come to an end and the Fed is on track to raise its rates by end of 2015, the following macroeconomic changes are certain. First, the rising benchmark interest rates would not be able to lure U.S. residents into increased borrowing for mortgaging cars and houses. Additionally, when interest rates rise, the investors who have invested for short-term gains in countries like India will move their investments for higher yields to the U.S. This would put a sudden strain on the Indian rupee (INR). Hence, the net result of rising rates in the U.S. with the present monetary policy would be a poor domestic consumer demand in U.S. from decreased borrowing, and a strain on economies of developing countries like India through rising inflation. These changes would cause a rise in value of the USD and depreciate the INR. Since one year of coming to power, although Prime Minister Narendra Modi has announced business-friendly policies, the INR has depreciated from Rs. 58 per USD to Rs. 67 per USD.

The net result of rising interest rates in the U.S. would be rising inflation in India as investors looking for better gains would rush to the U.S. for higher returns. The largest withholder of U.S. forex, viz. China, has signed currency swap deals with its major trading partners and performs transactions in Yuan instead of USD. Hence, although the USD will rise from foreign investments, the U.S. will not be able to reduce its trade deficits through exports as U.S. manufactured goods would become expensive in international markets. As there will be no major buyers for U.S. debt, due to bypassing of USD by major withholders of U.S. forex like China and Russia, the only way forward for the Fed is to reform its current monetary policy so that wages keep track with employee productivity, thereby reducing U.S. budget deficits. The U.S. has also recently imposed huge tariffs on solar goods from China and Taiwan to boost its domestic manufacturing. Eventually, the U.S. will also have to impose tariffs on all foreign goods entering the U.S., to eliminate its trade deficits and revive its domestic manufacturing industry. Without reforming its trade and monetary policies to reduce its trade and budget deficits, any rise in value of the USD with a rise in the Fed's benchmark interest rates would result in an increase in the U.S. twin deficits, which would also cause an increase in supply of goods to the economy, which is suffering from poor economic demand. This would cause a crash in profits of those corporations when their manufactured goods remain unsold, thereby also crashing the U.S. stock market.

Taking these macroeconomic changes into consideration, India has the following things to worry about for its Make in India plan. The plan to lure foreign investors to India to make India a global semiconductor manufacturing hub like China could fail, if any of these investors are looking for short-term gains, as semiconductor investments are long-term strategic investments. These investments pay off for any country over a long-term, they are not short-term investments that yield a quick return like the financial sector of today's economy. Hence, just like TSMC Inc. gets its financial backing from the government of Taiwan, Samsung Inc. gets its financial backing from the government of South Korea, and Globalfoundries Inc. gets backing from the government of Abu Dhabi, the upcoming Indian semiconductor fabs should also be sponsored with the backing of the government of India to make these capital-intensive investments sustainable. This approach would minimize any chance of these capital-intensive investments becoming unsustainable when investors move their investments out of India for their short-term gains due to the rise in U.S. interest rates. Fabs in India, backed by state or central government, could license manufacturing technology from developed economies and begin high-tech manufacturing in India.

These huge capital investments in semiconductor wafer fabs can become sustainable only if there is a solid economic demand for these semiconductor wafers in Indian electronics industry, because the economic demand is presently slowing in developing economies, too. However, if Indian government policies do not encourage consumption of domestic manufactured products and hence if the import of foreign manufactured goods continues due to India's free-trade policies, the trade deficits of India will continue to soar. The government recently passed some strict guidelines to all ministries, asking them to give preference to domestically manufactured electronic products. This is a positive step forward aimed at boosting electronics production as part of Prime Minister Narendra Modi's Make in India drive. If trade deficits are allowed to soar, they will put a further strain on the already troubled INR. Additionally, the products manufactured by Indian fabs will get consumed domestically only if the wages of Indian citizens keep track with their productivity. This free-market monetary policy based on the theory of mass capitalism would also ensure a robust consumer demand for electronic goods in order to keep the fabs in operation 24 × 7, thereby reducing the idle time of tools. Hence, Make in India needs to jump-start as Make in India for India. The potential of the fabless semiconductor ecosystem in its ability to grow small businesses should be adopted by the Make in India movement. In this process, several macroeconomic reforms should also be advocated, because an absence of these reforms have made the fabless semiconductor ecosystem unsustainable for the U.S., contributing to its twin (trade and budget) deficits.

To ensure a good RoIs by being able to create domestic jobs, and to minimize the job loses in economic downturns, a three-tier fabless semiconductor business model with decentralized supply chains, as explained in Chapter 5, should be adopted by the Indian semiconductor industry. This would usher in a competitive free-market balanced economy and help the Indian economy transition from a developing economy to a developed economy. The trade and budget deficits would be eliminated and ever-increasing huge capital investments would make this highly capital-intensive business not just sustainable, but also very profitable. This is a free-market approach to make the investments in semiconductor fabs a success so that India can keep its 2020 projected $400 billion account deficits, resulting from imported electronics, under control and ensure that its semiconductor manufacturing sector is able to keep track with the progress of Moore's Law.

Economy also moves in a systaltic fashion and never in a straight line. Due to this systaltic motion, internal clash and cohesion takes place, giving rise to economic cycles. The ups and downs of socio-economic life in different phases are sure to take place due to this systaltic principle. Having a balanced economic model will also eliminate the problem of unemployment in economic downturns. In this way, excess government spending in economic downturns can be eliminated, achieving a true free-market economic model for the Indian semiconductor industry.

10.4 A CROUCHING TIGER AND SLACKENING DRAGON TEACH A ROARING LION MACROECONOMICS

Don't be deceived by the title for this section. We are not going to talk about some *Jungle Book* story. This story, rather, is about three countries: one of them plans on making investments in semiconductor manufacturing and the others have experienced economic problems with such investments. History teaches us to learn from past mistakes and plan for a better future, but it seems that, sometimes, businesses ignore important historical facts, setting a course for themselves that ends in disaster.

To begin the story, let me introduce you to our three characters: a Lion, a Tiger and a Dragon.

10.4.1 WHO IS THE LION?

Make in India is a lion's step, said Prime Minister Narendra Modi, after launching the logo of his ambitious campaign to attract companies to India. The logo is the silhouette of a lion

Figure 10.1: Logo of Make in India Lion. Courtesy of `www.makeinindia.com`.

on the prowl, made entirely of cogs, symbolizing manufacturing, strength, and national pride. The national emblem of India, Ashok Chakra, also has four lions. In Indian folklore, the lion denotes the attainment of enlightenment, in addition to representing power, courage, pride, and confidence.

10.4.2 WHO IS THE TIGER?

Celtic Tiger (Irish: An Tíogar Ceilteach) refers to the economy of the Republic of Ireland between 1995 and 2000, a period of rapid real economic growth fuelled by foreign direct investment. The Irish economy expanded at an average rate of 9.4% between 1995 and 2000 and continued to grow at an average rate of 5.9% during the following decade, until 2008, when it fell into recession.

10.4.3 WHO IS THE DRAGON?

Chinese dragons are legendary creatures in Chinese mythology and Chinese folklore. Chinese dragons traditionally symbolize potent and auspicious powers, particularly control over water, rainfall, hurricanes, and floods. The dragon is also a symbol of power, strength, and good luck for worthy people. With this, the Emperor of China usually used the dragon as a symbol of his imperial power and strength.

10.4.4 WHAT ARE THE KING OF THE JUNGLE'S UPCOMING PLANS?

The idea behind the Make in India campaign is to promote India as a manufacturing hub and attract foreign investment in order to increase the share of manufacturing from 16% of GDP to 25%. Despite efforts made by the United Progressive Alliance (UPA) government to attract

Figure 10.2: The Celtic Tiger of Ireland.

more foreign direct investment (FDI) in several sectors, India's Foreign Direct Investments (FDI) inflows in 2013–2014 at $24.3 billion was just marginally higher than the previous year and much lower than the $31.5 billion inflow in 2011. The government of India also tried to raise funds through various initiatives, including the mega investors' summit that was launched in February

Figure 10.3: The Chinese Dragon.

2015 to attract global capital and give a push to Prime Minister Narendra Modi's signature Make in India program. Besides, the prime minister has traveled across the globe to bring investments of foreign capital into India.

Basically, what foreign investors want from their investments is to be able to reap high RoIs. Hence, the MNCs coming to India also look for low corporate tax rates. Additionally, in order to earn a high return on their investment, the MNCs also prefer to pay low wages as compared to what they have to pay to their workforce in developed economies. It is to be noted that tax cuts given to MNCs to lure them to India would contribute to India's budget deficit. Furthermore, as explained in *Mass Capitalism: A Blueprint for Economic Revival*, when the real wage fails to catch up with labor productivity, consumer debt rises sharply to prevent layoffs.

So, if the MNCs pay lower wages while extracting high productivity from India's labor force, consumer borrowing will soar. Much of this borrowing will be in terms of loans obtained on credit cards, with the borrowers paying as much as a 40% annual interest rate to Indian banks. Thus, the benefits from the FDI inflows are not as clear-cut as is traditionally claimed. Such investments do promote manufacturing, but they also contribute to budget deficits and consumer debt. In order to understand the macroeconomic problems that arise when wages trail productivity, let us observe the experience of the Celtic Tiger and the Chinese Dragon.

10.4.5 ECONOMIC BOOM TO BUST FOR THE CELTIC TIGER

During the 1990s, the rise of the Irish economy was one of the most remarkable post-war industrial phenomena. The key to economic growth was the massive inflow of FDI, as the MNCs sought to take advantage of Ireland's location and a young, well-educated labor force, along with corporate tax rates as low as 12.5%. The Irish economy grew at an annual rate of 10% and many foreigners immigrated to Ireland to take advantage of its booming economy. However, as wages failed to catch up with productivity, the technology-based, export-led growth turned into a bubble economy around 2002. The bubble burst in 2008, following the bankruptcy of Lehman Brothers in the USA.

Under strict guidelines enforced by the International Monetary Fund (IMF) and the European Union (EU), Ireland had to cut its budget deficit. However, the Irish national debt has continued to soar and unemployment remains stubbornly high at around 14%. Overall, the FDI experience for Ireland has been unpleasant in the long run. There was great prosperity in the short run, but now the nation is paying a big price.

10.4.6 THE SLOWING DRAGON OF CHINA

The HSBC Purchasing Manager's Index (PMI) for May 2015 indicated that China's factory activity contracted for a third straight month and output shrank at the fastest rate in just over a year. Because of low wages and high productivity, China has attracted vast inflows of FDI over the years. However, the suppression of wages in China could not continue forever. Eventually, the country had to raise wages, which has forced the United States and other MNCs to look for alternate manufacturing facilities to lower manufacturing costs.

China's growth has slowed considerably, and it is not clear if the nation will ever return to its spectacular growth again. Clearly, when wages trail productivity, the end result is either a weaker economy or rising unemployment, which is now the case in China.

10.4.7 WHAT LESSONS CAN THE LION LEARN FROM THE EXPERIENCES OF THE TIGER AND THE DRAGON?

Learning from the experience of the Celtic Tiger and the Chinese Dragon, the Make in India Lion should focus on boosting domestic demand by implementing free-market policies. This would ensure that wages keep track with productivity. *Mass Capitalism: A Blueprint for Economic Revival* provides a step-by-step approach to make this possible for the high-tech semiconductor manufacturing sector.

In this way, wage growth would lead to demand growth, which in turn would create entrepreneurship opportunities. Through this approach, Make in India would flourish irrespective of poor demand in developed or other economies. Free-market reforms that generate competition among various industries would also have to be carried out by other nations in order to come out of the present-day quagmire.

10.4.8   MAKE IN INDIA COULD COLLAPSE INDIAN ECONOMY

Ten months ago, Indian Prime Minister Narendra Modi launched his signature Make In India initiative to transform India into a global manufacturing hub. Less than a month later, I released the first book in United States to explore the macroeconomics of global semiconductor business entitled *Mass Capitalism: A Blueprint for Economic Revival*.

With a patriotic fervor and wanting to help the new government of India in its ambitious initiative, I arrived at my home country India and presented my ideas, in the form of a paper titled "Global Economic Crisis and Mass Capitalism as Blueprint for Economic Revival," to the Indian government's think tank in Mumbai in December 2014. I explained why the U.S. economy is headed for an economic collapse akin to 2008 due to crony capitalism resulting from a growing gap between wages and productivity in their economy. I also stressed that for Make in India to become successful, and to create a sustainable economy in India, the wages of the Indian workforce must catch up with worker productivity.

As I put forth in my book, I believe that globalization policies have resulted in a loss of dominance of the U.S. semiconductor industry due to the free-trade policies followed by U.S.-based multi-national corporations post World War II. These deceptive trade policies have also resulted in huge unemployment in the United States.

Even a country like China, which has received huge amounts of foreign investments from the United States, has experienced a spectacular economic boom. However, the absence of true free markets, a result of the state capitalism in China, has resulted in the gap between wages and productivity. China has cleverly controlled its exchange rates and avoided depreciation of USD by creating artificial demand for Chinese Yuan, in spite of huge trade deficits with the U.S. In this way, China has not only accumulated trillions of USD in its forex reserves but has also outsourced huge unemployment to the United States.

At the same time, since the wages of the Chinese work force have failed to keep up with worker productivity, China's stock markets started crashing earlier this summer. The trajectory is very similar to the economic crash that the United States experienced in 2008. The Chinese government has tried intervening in its economy but it does not seem to have helped, and the crash of Chinese stock markets continues.

When it comes to the Make in India initiative, not only has the government of India not endorsed any free-market policies, its economic policy makers have also not learned any lessons from the events happening across the world. Recently, there has been an attempt by the Finance Ministry to jeopardize the autonomy of the Reserve Bank of India (RBI). The Financial Sector Legislative Reforms Commission (FSLRC) published a first draft of recommendations, which gives the center the right to appoint four out of seven monetary policy committee members, and takes away the veto power of the RBI governor. I believe that this move has come into effect after reluctance on the part of the RBI's competent governor, Raghuram Rajan, to reduce RBI's repo-rates significantly, in spite of a push by the finance minister of India, Arun Jaitley. Rajan is much more worried about the impact of a hike in U.S. benchmark interest rates on the Indian economy, and hence does not want to lower repo-rates significantly.

In order to create a free-market economy, the central bank of any country should have two key characteristics:

1. Be under the control of a democratically elected government, and

2. Follow a monetary policy so that wages keep track with employee productivity.

Both of these have to be satisfied to establish a true free-market economy in India. As the Indian government does seem to be working towards establishing a true free-market economy, where wages catch up with productivity, the only option for the government to show economic growth (by means of a growth of the Indian stock markets) is to create a debt-based Indian economy similar to how U.S. ex-chairman of the Federal Reserve Alan Greenspan created a debt-based U.S. economy after 2001 by lowering the Fed's benchmark interest rates significantly, in order to lure more Americans into increased borrowing, and created a bubble economy. This resulted in a financial meltdown of the American economy in 2008 through the bursting of the unsustainable debt bubble.

Make in India should learn lessons of commonsense macroeconomics from the unpleasant experiences of the *Make in Ireland* and *Make in China* initiatives. Both Ireland and China experienced a short-term economic boom followed by a crash of their respective stock markets (due to the bursting of the debt bubble), in the absence of free markets, because of the growing gap between wages and productivity. Given a proven evidence of the failure of *Make in Ireland* and *Make in China* to create long-term stability and prosperity in either Ireland or China, a similar failure to create a free-market economy in India would have the same results. This would worsen the problems of poverty and unemployment in India, and prime minister Narendra Modi's approach to bring good days for India would, in fact, collapse the Indian economy and create worse days for the Indian economy in the near future.

The Make in India team can still create a robust economy for India by creating manufacturing jobs in the country and by learning from mistakes made by other countries. India can also transition to a developed economy by bringing out a sustainable economy when it implements a digital India revolution based on ideas presented in this book.

10.5   CONCLUSION

As India experiments with creative ways to solve its macroeconomic problems, and tries to make India a global manufacturing hub for high-tech manufacturing through its Make in India initiative, let the policy makers not forget what Albert Einstein once said, *"Insanity is doing the same thing over and over again and expecting different results."*

10.6 REFERENCES

[1] Anzuoni Mario, "U.S. Slap Hefty Duties on Solar Goods from China, Taiwan," Reuters, December 17, 2014. http://www.cnbc.com/id/102275081

[2] Batra, Ravi, End Unemployment Now: How to Eliminate Poverty, Debt and Joblessness Despite Congress, Palgrave Macmillan Trade, NY, 2015.

[3] Donovan, Donal, "Five important facts about the Irish economy," Oxford University Press's *Academic Insights for the Thinking World*. http://blog.oup.com/2013/09/five-facts-about-irish-economy/

[4] Mulay, Apek, *"Mass Capitalism: A Blueprint for Economic Revival,"* Book Publishers Network, Bothell, WA, 2014.

[5] "Make in India: PM to address global investors in Feb," *Hindustan Times*, Jan 10, 2015. http://theindiandiaspora.com/news-details/india-news/subprimary_news/make-in-india-pm-to-address-global-investors-in-feb.htm

[6] NDTV Correspondent, "India approves setting up of fab units at Yamuna Expressway and Prantij, Gujarat," *NDTV Gadgets*, February 15, 2014. http://gadgets.ndtv.com/others/news/india-approves-setting-up-of-fab-units-at-yamuna-expressway-and-prantij-gujarat-483854

[7] Yao, Kevin, "China May flash HSBC factory PMI shrinks for third month, more stimulus seen," Reuters–Beijing. http://www.reuters.com/article/2015/05/21/us-china-economy-pmi-hsbc-idUSKBN0O605620150521

CHAPTER 11

The Internet of Things (IoT) Revolution

11.1 WHAT IS IOT?

The internet of Things (IoT) is nothing but a network of several appliances and electronics that are used everyday along with sensors for turning on and off different devices which are connected to the Internet. This technology would assign a unique IP address to every appliance, making it uniquely identifiable on the network. Additionally, it would be possible to remotely control the operation of these appliances. The advantage of this approach being that it would lead to the next digital revolution on a massive scale. While the IoT certainly sounds like a great way forward, there are countless challenges when it comes to implementation and affordability to make it sustainable. In this chapter, we shall explore how the IoT revolution can become sustainable and lead to the next high-tech revolution by ensuring a sustainable macroeconomic progress. We shall also look into how the progress of Moore's Law can be sustained with this IoT revolution and why it is essential for the IoT to succeed for the growth of today's knowledge-based economy.

11.2 CASE STUDY–*GOOGLE NOW* AND IOT

When Google Inc. started as a company in 1998, it was a small company with great ideas. The visionary founders realized, at a very early stage, the importance of a great infrastructure on which to offer its search engine services. From its infancy, Google Inc. worked hard to set up its state-of-the-art infrastructure, involving huge costs and long working hours.

Today, Google Inc. is about to transform the IoT industry very similarly to how the semiconductor industry has been transformed from an integrated device model (IDM) to a fabless-foundry business model. The Cloud Platform under Google Now offers an opportunity for new start-ups to focus on innovations and invest a fraction of the costs that would be needed for owning a state-of-the-art infra structure. This is very similar to how the fabless-foundry model of the semiconductor industry led to the democratization of innovation by ushering the growth of several small fabless semiconductor businesses that focus on innovative circuit designs and transferring the manufacturing costs to their foundry partners. The fabless-foundry business turned out to be a win-win for both fabless semiconductor companies and their foundry partners, thus leading to a transformation of the global semiconductor industry.

Through the Cloud Platform, new software startups can make use of the state-of-the-art infrastructure of Google Now, Amazon Web Services, etc. and offer high-quality networking services, computational services, big data services, and other software services. The state-of-the-art infrastructure of Google Now also needs to be constantly upgraded with the progress of Moore's Law. Through the Google Now platform, the developers can gain access to a superior technology at a fraction of the costs of owning a state-of-the-art infra structure. Google Now is indeed a well-thought move by Google Inc. that benefits new start-ups in the IoT industry to benefit from falling prices of state-of-the-art infrastructure due to mass production in compliance with the progress of Moore's Law.

Today, cloud service providers like Google Now, Amazon Web Services, etc. should be aware of the importance of sustainable macroeconomic progress for sustaining the progress of Moore's Law and for robust growth of the IoT industrial sector. There should be both private companies like Google Inc., Amazon Inc., etc., as well as cloud service providers supported by local government, to have a healthy competition as well as to prevent the formation of monopolies in this industry. A balanced economy would also benefit the public cloud service providers, like Google, Amazon, etc., by ensuring a good ROI because of the high costs involved in upgrading their hardware infrastructure to provide a state-of-the-art infrastructure to their customers, who are the small- and medium-size software service providers.

These software service providers can in turn focus on providing innovative services leading to rapid innovation and growth of the IoT industry. This would not only benefit the end user of services but would also save the costs of huge investments in state-of-the-art infrastructure for the service providers. By benefiting both producers and consumers in the industry it would usher in an economic democracy in the high-tech industry. A decentralized supply chain between the providers of these software services would also prevent any mergers and acquisitions (M&As), which result in the growth of internet monopolies, which hurt the overall macroeconomic growth due to the loss of free markets. Hence, for robust growth of the IoT sector as well as to comply with macroeconomic cycles of nature, a three-tier business model is recommended, based on the theory of mass capitalism, as explained in Chapter 5.

11.3   NET NEUTRALITY PRESERVES A FREE-MARKET ECONOMY FOR IOT SUCCESS

The real job creators in any free-market economy are not only producers but also consumers. Monopoly capitalism has resulted in poor economic demand due to reduced consumer purchasing power in the U.S. and global economies. In Chapter 6, I discussed the possibility of an early demise for Moore's Law due to monopoly capitalism, as opposed to a natural demise due to physics. In Chapter 8, "The Macroeconomics of 450 mm Wafers," I have explained how the loss of a free-market economy due to monopoly capitalism produced a poor RoI in the transition to 450 mm silicon wafers for the global semiconductor industry, bringing this profitable business to a standstill.

At the 2014 Web Summit in Dublin, Ireland, industry leaders came together to discuss the future of the IoT. The global semiconductor industry is now interested in the IoT mainly because the Internet has flourished at a greater level than the voice telephony industry, due to an absence of government-sanctioned monopolies. In Chapter 9, "Moore's Law Beyond 50," I make a case that, even though voice-over Internet protocol (VoIP) requires the same bandwidth to send voice data as Internet packets when compared to voice packets transferred over standard telephony infrastructure, the prices that consumers pay for VoIP services are much lower than what they pay using standard telephony infrastructure.

The real reason voice telephony infrastructure prices have not fallen, in spite of the rapid market growth of cell phones and smartphones, can be traced to government-approved monopolies. Political corruption has enabled crony capitalists to influence the decisions of the United Nations' International Telecommunications Union (ITU) and prevent telecommunications prices from falling, and thereby benefiting consumers, in spite of the relentless progress of Moore's Law. Since VoIP did not make use of an infrastructure from the old telephony system, it avoided monopolies and fueled a revolutionary growth of the Internet. Now monopoly capitalism in the semiconductor industry is bringing this capital-intensive business to a standstill. Very few players dominate this industry. Meanwhile, crony capitalists are also trying to establish Internet monopolies through their lobbying efforts because of the upcoming IoT revolution.

As a result of these lobbying efforts, in January 2014, the U.S. Court of Appeals, the second-highest court in the U.S., knocked down the Federal Communication Commission's longstanding regulation requiring "network neutrality." Without network neutrality, the growth of the IoT industry will likely come to a standstill due to poor consumer demand, just like our global semiconductor industry. Net neutrality is crucial in a true free-market enterprise system. In such a system, the size of government is small and the power of businesses remains in check. A free-market economy ensures a healthy competition among businesses, thereby bringing down the costs for end consumers.

A world without net neutrality allows formation of Internet monopolies, to enable time-sensitive packets, making it possible for big money to pay more to have their internet packets placed ahead of the queue and be given preferential treatment, even if the content is the same as other sites. In this climate, if net neutrality is not protected, small businesses in all industries that rely on the Internet will be adversely affected. Large corporations with deep pockets could ultimately access data at faster speeds than their competitors with less capital. Small businesses would start experiencing slower speeds and higher prices. Consumer prices could also be driven up due to lack of competition. What's more, we would not have the next Google Inc., Facebook Inc., Twitter Inc., etc., which, believe it or not, were once small businesses. The Internet would no longer be a free and fair medium; big money would win the game. If we want a healthy free-market enterprise, we need to foster a healthy economic environment that allows growth of talented new entrepreneurs. Net neutrality is key to this. The alternative undercuts innovative services and growth of small businesses, which would also hurt the growth of the IoT revolution.

Since the United States does not have free-market capitalism anymore, as it had in the 1950s–1970s, government intervention offers one way to stop the growth of Internet monopolies. Hence, President Obama is absolutely correct in calling on the FCC to impose "the strongest possible rules" to protect net neutrality—prohibiting Internet service providers from favoring some lawful content over others, and thus avoiding the creation of Internet industry monopolies that would stagnate the growth of the IoT sector. It is possible to have minimal government intervention in the economy and a self-regulation of the economy only by means of implementing a free-market economic system. Free-market economic reforms would also result in an exponential growth for the IoT sector. Letting employee wages automatically catch up with their productivity would boost consumer demand in the economy.

Employee-owned corporations with a neo-cooperative management would also ensure that all the people in an economy are real stakeholders of that economy, thereby establishing an economic democracy. This new system would create plenty of jobs and put an end to the debate about having only a few people pay taxes while the rest live on unemployment benefits. It would also eliminate political corruption in the U.S. democracy, as majority shareholders of the corporations (i.e., employees) would not allow passage of laws that would hurt their own interests. An economic democracy would add value to the political democracy and make it a stakeholder democracy, which would truly work as a government of the people, for the people, and by the people. In this way, *mass-capitalism*-based reforms offer a free-market solution to achieve net neutrality with minimal government intervention, through a smaller government, and with lower taxes on individuals for success of the IoT revolution.

11.4   *MASS CAPITALISM* PAVES A PATH FOR THE IOT REVOLUTION

The upcoming IoT revolution hopes to extend the end node far beyond the human-centric world to encompass specialized devices with human-accessible interfaces, such as smart home thermostats and blood pressure monitors, and even those that lack human interfaces altogether, including industrial sensors, network-connected cameras, and traditional embedded systems. As the IoT grows, the need for real-time scalability to handle dynamic traffic bursts also will increase. Each of those IoT end nodes requires connectivity, processing and storage, some local, some in the cloud. This means scalability, reliability, security, compliance, and application elasticity to adapt to dynamic requirements and ever-changing workloads.

A three-tier business model would lead to an increase in consumer purchasing power in the economy, thereby increasing the domestic consumer demand. Only when the consumer demand rises through growth in wages, as compared to luring consumers into debt, can the economic demand remain sustainable. When such free-market economic reforms become a reality in the United States as well as global economy, it would lead to an increased demand for more IoT products, like thousands of medical devices in hospitals, smart utility meters, GPS-based location systems, fitness trackers, toll readers, motion detector security cameras, smoke detectors, embedded systems, etc.

As shown in Figure 11.1 below, in order to ensure that the IoT revolution is sustainable and leads to a balanced economic growth, where every end node on the Internet, private network, and servers for central systems are able to connect to the private or public cloud, a three-tier business model as presented in Chapter 5 is highly recommended. In this model, the upper industrial tier would include the servers that need huge capital investments and constant upgrade, which constitutes public and private clouds like Google Now, Amazon Web Services, etc. The middle industrial tier (neo-cooperative corporations) includes the public and private broadband Internet providers, networking service providers, computational service providers, and Big Data service providers, as well as other software service providers. The lower industrial tier would be the small businesses that can be connected via a network or be individually addressable networks (which could be a LAN, PAN, body area network, etc.).

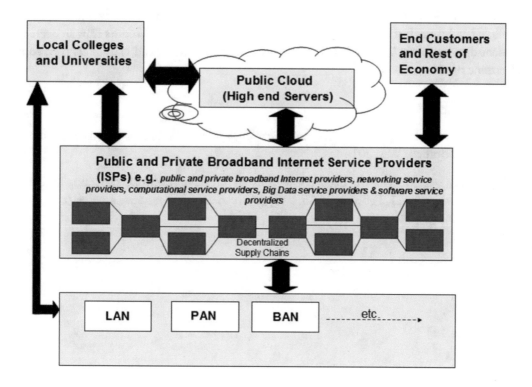

Figure 11.1: A three-tier business model for robust growth of the IoT sector.

Low unemployment is key to economic stability and robust growth of domestic consumer demand. Only by means of establishing a true free-market economy, can consumer demand grow in a sustainable way so that the producer of goods keeps investing in better products that cater to the needs of consumers. The IoT specifications also call for lower-power devices so that it becomes feasible to have several devices talk to each other in close proximity, leading to a strong focus on local economic development. In this regard, I have already provided an in-depth analysis in Chapter 7 of the ability of decentralized supply chains to bring about a robust growth of local economies, which would further boost the success of the IoT revolution, thereby making the semiconductor supply chains more efficient.

11.5 CONCLUSION

In this way, *mass-capitalism*-based free-market reforms would result in the IoT revolution becoming a grand success story for global high-tech industries and also for the global economy. Additionally, when the consumer purchasing power keeps growing, there would be an incentive for semiconductor foundries to make investments in the progress of Moore's Law, as well as transition to 450 mm diameter silicon wafers. In this way, mass capitalism would enable not just the IoT revolution, but it would also ensure sustainable progress of Moore's Law to overcome its present day economic limits when growing economic demand would call for its own supply of electronic products.

11.6 REFERENCES

[1] Mulay, Apek, *Mass Capitalism: A Blueprint for Economic Revival*, Book Publishers Network, Bothell, WA, 2014.

[2] Mulay, Apek, "Do we need regulations to preserve 'network neutrality'?" *The Costco Connection*. p. 21, February 2015. http://www.costcoconnection.com/connection/201502?pg=24#pg24

CHAPTER 12

An Engagement with Semiconductor Industry Thought Leaders about the Future of the Semiconductor Industry

12.1 INTRODUCTION

Throughout history, the semiconductor industry has continued to change and evolve—and we have looked at industry pundits to point us the way. That conversation is a critical part of forging our new path. Recently, I read a volume titled *Fabless: The Transformation of the Semiconductor Industry*, by author Daniel Nenni and contributor Paul McLellan. In the final chapter, the authors polled a variety of industry names with the question: "What's next for the semiconductor industry?" I wanted to engage with these thought leaders and share thoughts I've developed and written about on macro-economic principles based on the theory of mass capitalism for sustaining the progress of Moore's Law to ensure a success of the IoT revolution. We shall conclude this chapter after analyzing input from Chenming Hu, Robert Maire, Dan G. Hutcheson, Bijan Davari and Mark Bohr.

12.2 A VIRTUAL SIT-DOWN WITH INDUSTRY LEADERS PART I

In this virtual sit-down, I'll be addressing the thoughts of several semiconductor leaders, including Simon Segars, CEO of ARM; Aart de Geus, Chairman and CEO of Synopsys; Lip-Bu Tan, Chairman and CEO of Cadence Design Systems; Dr. Ajoy Bose, CEO, at Atrenta.

Segars makes in interesting statement in the book, when he says, "*Without the seamless design-to-end-user supply chain that has evolved through the growth in the fabless industry, our vision would remain a dream.*" He believes that the movement toward fabless semiconductor companies has democratized innovation. I am in agreement with Segars, but would add that the growth of small businesses should be credited with the success of the fabless semiconductor industry. We are still striving for democratized innovation, which will only be achieved when we have a functioning economic democracy.

Meanwhile, Aart de Geus puts forth that semiconductor innovation would be most visible through utilization. I believe that this utilization must be of a progressive nature for innovations to benefit the masses. Only by means of benefiting all will it be possible to visualize the true benefits of innovation. For this to materialize, a free-market economy has to be established. He also points to the enabling of "better" and "sooner" innovation (rather than simply cheaper) as a desirable means of keeping pace with the speed of innovation in the semiconductor food chain. I would respond that the best, and perhaps only, sustainable path to this sort of innovation is through a decentralized supply chain.

Cadence's Tan calls for deep collaboration in order to reach higher levels of innovation and financial success in the supply chain. He says that no one in this ecosystem can succeed alone, but we can all succeed together. I am of the opinion that this collaboration can be achieved only when all participants in this fabless semiconductor ecosystem have a stake in the success of the entire ecosystem. Hence, there has to be a change in corporate business managerial systems to ensure accountability and shared growth and success with all participants, which includes every employee who is a part of the fabless semiconductor ecosystem.

Finally, Dr. Bose of Atrenta suggests that low-margin demand may be a simple reality in the semiconductor market, the only possible exceptions being those elite few who are able to move up the chain to offer more total ownership of systems and services. He further says that there would be a continued drive to move design to low-cost regions, and the clock is running out for any suppliers unwilling to adapt. I would say that Dr. Bose is right. The cause for low-margin demand is lost consumer purchasing power due to U.S. macro-economic policies. The Fed's monetary policies since 1970 have resulted in crony capitalism, which has transformed the U.S. economy from a free-market enterprise system to monopoly capitalism. This has resulted in total ownership by an elite few. Following policies of globalization with free trade, which drive down the costs of design by moving design operations to low-cost regions, would further exacerbate the problem of low marginal demand because of trade deficits that would be added to the U.S. economy.

12.3 A VIRTUAL SIT-DOWN WITH INDUSTRY LEADERS PART II

As discussed, innovation and collaboration will be key tools for semiconductor makers as they drive the industry forward. In this virtual sit-down, I'll be addressing the thoughts of other semiconductor leaders including Jack Harding, President and CEO of eSilicon; Kathryn Kranen, CEO of Jasper Design Automation; David Halliday, CEO of Silvaco and Dr. John Tanner, CEO of Tanner EDA.

Today, we're testing the levels of optimism of a handful of semiconductor thought leaders. Jack Harding, for one, is optimistic. In the book, he credits the semiconductor industry with some of the world's boldest innovations, noting that the industry has solved *"countless problems with uncanny predictability, and this will not stop.*" Harding is very optimistic that the semiconductor industry will progress to overcome the economic barriers of Moore's Law.

Kathryn Kranen, however, calls the semiconductor to a higher bar. She says that, to date, the semiconductor or electronic design automation (EDA) ecosystem has not fully taken advantage of the connected and collaborative world it has helped to create. I agree with Kranen's assessment. The primary issue is that the current fabless-foundry business model does not ensure a complete collaboration among all players and contributors in this ecosystem. Some major managerial reforms are needed to ensure that we in the industry fully leverage the strengths offered by the connected, collaborative world that has been created.

David Halliday, credits the customer with industry growth and says that a new and major consumer revolution is necessary to bring forth the next Golden Age of semiconductors. I would say that the last Golden Age of free-market economy occurred in the U.S. from 1950 to 1970. To recapture that time, we need a major consumer revolution in terms of increasing the consumer purchasing power through major reforms in the U.S. as well as global economy.

Finally, to reach this Golden Age, connections among industry players are not enough, according to Dr. John Tanner: "*Inter-disciplinary, intra-industry, and cross-industry connections are required. As technological and business issues grow more complex, we'll need to draw cross-domain expertise capability into the fold.*" I believe that the semiconductor industry's fabless-foundry business model needs to undergo some reforms in order to increase collaboration and cooperation across different disciplines. I have made recommendations and proposed a new three-tier business model for the fabless semiconductor industry that would not just ensure a great collaboration, but would also enhance cooperation in inter-disciplinary, intra-industry, and cross-industry avenues.

12.4 A VIRTUAL SIT-DOWN WITH INDUSTRY LEADERS PART III

The semiconductor industry has a history of evolution and revolution, both from a technological and a business point of view. Yet there is little agreement among industry leaders on how to forge a path forward in this moment in time. In this virtual sit-down, I'll be addressing the thoughts of other semiconductor leaders including Jodi Shelton, Co-Founder and President of Global Semiconductor Alliance; Grant Pierce, CEO of Sonics; Sanjiv Kaul, CEO of Calypto Design Systems and Hossein Yassaie, CEO of Imagination Technologies.

Today, the business landscape for semiconductor makers is changing. The industry's top CEOs are concerned with the blurring of traditional demarcations between supplier, partner, customer, and chip designers, Jodi Shelton told the book's authors. She added that further consolidation, not only in the semiconductor industry, but also within the related ecosystem of partners and customers, is also of note. I would say that certainly, CEOs should be concerned. It is to be noted that this consolidation hurts innovation and also affects consumer demand. Some consolidation goes against the principles of a free-market economy. It should be noted that U.S.

capitalism needs major reforms toward a free-market economy or else the U.S. economy will see an early demise of Moore's Law for this great industry.

Meanwhile, Grant Pierce pointed to the difficulty of predicting market change in the face of device sizes that approach the atomic level. "*He can predict that innovation will continue,*" he said in the book. "*Innovation will answer the call from market. The semiconductor industry is that good.*"

It is true that all semiconductor industry professionals are concerned about the future of the semiconductor industry as it approaches the physical limits of Moore's Law. Just like Pierce, I also believe that innovation will answer the market's call. However, for innovation to answer the call, there has to be a free-market economy, which leads to a robust growth of small businesses. Only when there is a robust growth of small businesses can there be a growth of new start-ups that in turn leads to the birth of new ideas and innovations that better the lives of human beings. For this innovation to answer the call from the market, the semiconductor industry should help transform the present U.S. economy from crony capitalism to a free-market economy. By doing so, the semiconductor industry would prove that it is "*that good.*"

Of course, a perennial concern for the industry is profits. Sanjiv Kaul believes that the ability to generate a leadership position or even to remain relevant becomes questionable in the face of the capital-intensive nature of the business. He points to this reality as a factor in industry consolidation. He adds, "*The interesting question is which companies will come out the winners in this round of battles for supremacy?*"

There's no denying the capital-intensive nature of the business. However, the present colonization of the publicly traded companies by Wall Street has been responsible for the irrational distribution of profits resulting in increasing mergers and acquisitions (M&As), which kills innovation.

M&As invariably lead to larger and less efficient organizations. Many times these business deals result in layoffs, which further depreciates consumer demand in the economy. To answer Kaul's question, I would say that no company would come out a winner in this round of battles for supremacy, because supply and demand are like two wings of a bird and a one-winged bird cannot fly. To put an end to consolidation in the industry, the semiconductor industry has to undertake reforms to ensure a rational distribution of profits across the board and make its employees, rather than outside Wall Street investors, majority shareholders of the corporation.

Hossein Yassaie, meanwhile, calls for a laser focus on customer wants and needs and a mandate on engineering to fulfill those needs as solutions. Yassaie's statement points to an important fact of economics, which is unfortunately being ignored by the semiconductor industry at large.

Consumer demand for products forces the industry to manufacture products to meet those demands. Hence, the supply of goods and the economic demand for them takes place with the cooperative action of producers and consumers. An oversupply of goods and weak demand results in layoffs at the manufacturing facility. Consumer demand invariably drives the manufacturer to increase manufacturing levels. Good consumer demand acts as an engine of economic growth in a free-market economy, and industry produces goods towards fulfilling those needs.

12.5   A VIRTUAL SIT-DOWN WITH INDUSTRY LEADERS PART IV

The newest trends in the industry, from new design tools to new product categories, promise to give the semiconductor industry a lift. In this virtual sit-down, I'll be addressing the thoughts of other semiconductor leaders including Srinath Anantharaman, CEO of Cliosoft; Mike Jamiolkowski, CEO of Conventor and Joe Sawicki, Vice President and General Manager of Mentor Graphics.

The newest generation of electronic design automation (EDA) tools has increased designer productivity in order to keep up with Moore's Law, Srinath Anantharaman, said in the book. At the other side of the equation, growing teams invite inefficiency.

To his point, certainly the productivity of designers has been steadily growing with new technological innovations. This employee productivity increases the supply of goods into the economy. In the case of EDA, productivity increases stem mainly from new and innovative circuit designs for electronic products. However, in order to increase the economic demand in proportion to increased supply from rises in productivity, CEOs must ensure that the wages of the designers who are highly productive grow with their productivity. Smaller organizations increase the efficiency of employees. Higher efficiency in larger organization can be achieved by decentralization of different engineering teams so that they are each about the right size so that the management can ensure that wages keep track with productivity.

Of course, historically and into the future, semiconductor makers are betting on new technologies and processes to stay ahead of the pack. Mike Jamiolkowski, estimates that $50–$100 million is spent every two three months on new process technology development that focuses on iterative trial-and-error cycles of learning, using wafer-based experiments in fab for 3D IC process development.

Based on Jamiolkowski's estimates, when the semiconductor industry spends so much money on new process technology development, it is advisable for the industry to also ensure that the high-tech manufactured products are in high demand in the economy. What is the use of making huge investments when there won't be a significant demand to give a good return on the investments? This points us to an important fact that, for making the continuous capital investments sustainable, the semiconductor industry, like other industries, should have a business model that boosts consumers' purchasing power so that there is a continued demand for better products. This would force the manufacturer to keep investing in newer technologies and make these investments sustainable.

Another top exec, Joe Sawicki is betting on the emerging IoT market to drive an increase in productivity that will dwarf that gained by PCs and mobiles. "*Semiconductor systems enabling this trend will need to respond to difficult cost, size, and constraints to drive real ubiquity,*" he said in the book. I do not doubt that IoT would drive an increase in productivity as long as net neutrality is maintained. However, if wages of employees in the semiconductor industry do not catch up with their productivity, the overall economy will continue on its unbalanced path, which will result in diminished consumer purchasing power that could translate into poor demand for consumer electronics in the IoT market.

12.6   A VIRTUAL SIT-DOWN WITH INDUSTRY LEADERS PART V

Foundries continue to face increased competition in the semiconductor industry, demanding new approaches in order to achieve success. In this virtual sit-down, I'll be addressing the thoughts of other semiconductor leaders including Subi Kengeri, Vice President of Advanced Technology Architecture at GlobalFoundries; Richard Goering, veteran EDA editor and Senior Manager of Technical Communications at Cadence Design Systems; Rich Goldman, Vice President of Corporate Marketing and Strategic Alliances at Synopsys and Semiwiki blogger Luke Miller.

Subi Kengeri told *Fabless* authors that the era of foundry 2.0 has arrived. He believes that only companies that are willing to collaborate early, deeply, and openly will survive and thrive in this emerging environment. I should applaud him because his thoughts met with action when GlobalFoundries announced that it would be collaborating with Samsung and utilizing Samsung's 14 nm manufacturing process. In order to ensure a better cooperative collaboration across the entire industry, there need to be reforms in existing business models and supply chains to achieve this greater cooperation over competition. Only a fabless-foundry business model can achieve a better cooperative collaboration, which would benefit both fabless businesses and their foundries.

Richard Goering concurs that deep and early collaboration among participants in IC design supply chains—including IP providers, EDA vendors, and foundries—is a must. "*As we grapple with problems of today, let's lay the groundwork for the bright ideas of tomorrow,*" he said. Mr. Goering's vision of supply-chain collaboration can be realized only through decentralized supply chains, which lead to an increased cooperation over competition between the entities in the supply chain. Only decentralized supply chains lead to a flourishing of individual entities. There is less to lose and more to gain by cooperating with other entities in the supply chain. I believe visionaries like Richard Goering would support laying the groundwork for establishing a decentralized supply chain.

Rich Goldman believes that keeping up with the pace of Moore's Law continues to get more and more difficult, both physically and economically. "*The economic challenges are forcing us to advance in three axes by developing new methodologies and experimenting [with] new materials,*" he said in the book. I appreciate Goldman pointing to twin challenges with progressing Moore's Law both physically and economically. While 3D ICs are an innovative way for scaling transistors, the scaling in the third dimension could prevent making our smartphones slick and slim for every generation, although it would continue the progress of Moore's Law.

The same also applies to developing new methodologies and experimenting with new materials to prolong the physical limits of Moore's Law. The economic challenges of Moore's Law could be overcome by some commonsense macroeconomic reforms. Only by having a free-market economy will it be possible to have a sustainable economic demand for advanced electronic products. When there is a sustainable demand for the latest and greatest electronic products, then the investments made by the semiconductor industry to overcome the physical limits of Moore's Law could be made sustainable, since they would create a high demand for the manufactured products.

Semiwiki blogger Luke Miller is optimistic about the industry's ability to meet the challenges ahead, saying, "*As we step out further, electron or quantum computing will unleash a technological world that would blow our minds, solving mysteries and problems once thought impossible.*" I support Luke and, in order to achieve that transition, the economic limits of Moore's Law also have to be overcome. I am sure, with some solutions which I have proposed in this book, those economic limits that were once thought of as impossible to conquer would also be easily overcome and eventually it will be physics that will lead to a demise of Moore's Law.

12.7 A VIRTUAL SIT-DOWN WITH INDUSTRY LEADERS PART VI

In high tech, we talk about innovation and disruptive technologies, but now it's time to talk about what these really look like in the semiconductor industry. In this virtual sit-down, I'll be addressing the thoughts of other semiconductor leaders including Semiwiki blogger Erik Esteve; Semiwiki blogger, Daniel Payne; Semiwiki bloggers and authors of the volume titled "*Fabless: The Transformation of the Semiconductor Industry,*" Paul McLellan and Daniel Nenni.

Erik Esteve believes that innovation has always been the driver for a highly dynamic, creative, and successful semiconductor industry. He hopes that industry actors will be wise and leave room for startups to emerge and innovation to flow. Such creativity would enable post-smartphone products of 2030 or later, keeping the semiconductor industry as dynamic and successful as it has been up to now, according to Esteve.

I would answer that a couple of things need to happen to support this vision. For robust growth of small startups, it is very important that U.S. capitalism undergoes major reforms leading to a free-market economy. Additionally, mergers and acquisitions form inefficient, large organizations and lead to layoffs when there is a merger between companies that do not have complementary portfolios. Hence, antitrust laws have to be strictly enforced in order to ensure free markets, and preserve innovation as well as sustain the growth of a highly dynamic, creative, and successful semiconductor industry.

Daniel Payne said that he hopes to see some new, disruptive nanotechnology emerge that would disrupt the status quo and incremental improvements. I would like to see Daniel Payne's vision turn into reality and believe that with good economic reforms and growth of startups, new technologies are sure to emerge that would disrupt the status quo and incremental improvements.

A good economic system would solve, not just current, but also future problems of the global semiconductor industry.

Paul McLellan is less optimistic. He predicts that the *"semiconductor industry is entering an era where economics that have driven semiconductor for five decades may be coming to an end... On the other hand, something may change, or maybe this time it really is different."*

McLellan is correct in his analysis that today's U.S. economy cannot sustain Moore's Law anymore, and it is indeed coming to an end due to economics. During these five decades of progress of Moore's Law, free-market capitalism has been transformed into monopoly capitalism. As long as economic reforms bring back those free markets, a brilliant future can be envisioned for the global semiconductor industry. If there is something that should change then it is the present form of U.S. capitalism.

Finally, Daniel Nenni weighs in himself, saying that the biggest challenge for the semiconductor industry to move forward is economic. Nenni believes the fabless semiconductor ecosystem will clear technical hurdles for physical limits. However, he says, we are under-estimating the increasing financial burden of modern semiconductor design and manufacturing. Nenni contends:

> *"As our industry matures, consolidation comes knocking and doors of innovation start closing. Today less than 5% of the workforce is engineering and science based. If you really want to know what will finally kill Moore's Law, the answer is economics absolutely."*

Nenni is correct that the increasing financial burden of modern semiconductor design and manufacturing is becoming too risky an investment for many businesses. He is also correct that very little of the workforce is involved in engineering and scientific advancements. However, I believe that mass-capitalism-based free-market reforms offer the only solution to these economic problems. This is the only approach in which wages of employees catch up with their productivity, establishing a free-market economy. This creates a robust economy while keeping the rising economic inequality in check, with minimal government intervention, which leads to a robust growth of small businesses.

With a decentralized supply chain, individual entities in the supply chain prosper and there is more cooperation than competition. This kind of cooperation leads to collaboration among businesses. By focusing on the growth of science, technology, engineering, and math (STEM) education, it is possible to increase the percentage of tomorrow's workforce in these fields. Some forward thinking and planning would solve even the economic hurdles to the progress of Moore's Law. In fact, the present physical and economic challenges to the progress of Moore's Law are a boon in terms of helping bring about reforms in the global economy that will lead to all around progress of all human beings. It gives us an opportunity to make capitalism work for everyone. Here, I would like to re-iterate what I have already explained in Chapter 6, that it will be physics, and not economics, which would eventually lead to the demise of Moore's Law.

12.8   CONCLUSION

As we conclude this chapter with an engagement with semiconductor industry thought leaders about the future of the semiconductor industry based on their inputs in the volume *Fabless: The Transformation of the Semiconductor Industry*, let me also take into consideration some other views expressed by Chenming Hu, Robert Maire, Dan G. Hutcheson, Bijan Davari and Mark Bohr in an article authored by Rick Merritt in *EE Times* entitled "Moore's Law Hits Middle Age—Tales from engineers who drove it forward."

Chenming Hu, professor at Berkeley, California, as well as former chief technologist at TSMC, said, "*No exponential growth can go without an end.*" He added, "*We are getting to the point where we are counting atoms, but I think Moore's Law end will come earlier than that.*" Hu said, "*In making things smaller in linear dimensions I believe we are not far from the end, so we will have to look for other ways forward.*"

I agree with Professor Hu about the necessity for new ideas to take Moore's Law forward. In Chapter 9 "Moore's Law beyond 50," I provided a free-market solution to continue this growth in the number of transistors per square inch of an IC by ensuring a sustainable consumption of electronics through the establishment of free-market economic policies.

Robert Maire, a semiconductor analyst, said, "*To date, the progress Moore's Law represents has not been limited to just ever faster and cheaper computers but an infinite number of new applications from communications and the Internet to smartphones and tablets. No other industry can claim similar far-reaching impact on the lives of so many people [in] less than a lifespan, more changes in the world can be traced back to the enabling power of the semiconductor industry than any other industry...More lives have been saved and fortunes impacted.*"

Robert Maire is precise in his analysis and hence it is very important for the semiconductor industry professionals to look for new ideas to sustain a progress of Moore's Law. Mass capitalism presents some out-of-the-box ideas for the sustainable progress of Moore's Law and also analyzes the macroeconomic policies that have resulted in an earlier demise of this law.

Dan G. Hutcheson, CEO at VLSI Research, said, "*The market value of the companies across the spectrum of technology driven by Moore's Law amounted to $13 trillion in 2014. Another way to put it is that one-fifth of the asset value in the world's economy would be wiped out if the integrated circuit had not been invented and Moore's Law never happened.*"

Dan points out the important role played by Moore's Law in the global economy. As long as the progress of this important law can be sustained, the global economy will continue to prosper from technological innovations, or else it will lose 20% of its share due to the demise of this law. In this regard, proper macroeconomic policies have become critical for the progress of this law. With a balanced free-market economic approach, there would not be any trade deficits or budget deficits. Besides, increasing consumer demand will also increase the sales of electronics and thus increase the market value of companies across the spectrum of technology driven by Moore's Law.

Bijan Davari, an IBM fellow, said, "*It won't be like traditional Moore's Law scaling in the future, it will be more of a combination of things. Performance came from frequency enhancements from megahertz to gigahertz, a thousand-fold progress in 20 years, and most of that came from device scaling. Now because that has almost come to a stop, we have to go to parallel computing with many cores and accelerators."*

While I agree with everything that Bijan Davari said, I would argue that, for Moore's Law to continue beyond 50 years, it needs to ensure that there is a sustainable consumption of manufactured electronics to sustain an increased demand for companies to make further investments for the technological progress of Moore's Law. The macroeconomic policies should provide an incentive to not just producers but also consumers in the economy.

Finally, Mark Bohr, a senior fellow in Intel's manufacturing group, said, *"What started out as an observation became a guide for us all that we felt we needed to follow and, if possible, faster than anyone else in the industry. Overall I don't think [Moore's Law] will end, it will evolve and change in terms of what we do. Our industry's ability to innovate and develop ICs of one sort or another will continue for a long time...exponentials are difficult to sustain, [so] the slope [of progress] may not be quite as steep ten years from today as now, but it will continue."*

Like Mark Bohr, I am equally optimistic about the progress of Moore's Law and believe that innovations will continue in the semiconductor industry. The upcoming IoT revolution offers an opportunity to exponentially grow the supply of consumer electronics to the economy. However, until the semiconductor industry realizes that the growth in consumer purchasing power is equally important, just as is the growth in the number of transistors per square inch of an IC, the progress of Moore's Law would be unsustainable. My forecast is that this progress will continue and mass capitalism will drive the progress of this law beyond its 50 years.

12.9 REFERENCES

[1] Merritt Rick, "Moore's Law Hits Middle Age- Tales from engineers who drove it forward," *EE Times*, April 14, 2015. `http://www.eetimes.com/document.asp?doc_id=1326336&page_number=1`

[2] Mulay, Apek, *Mass Capitalism: A Blueprint for Economic Revival*, Book Publishers Network, Bothell, WA, 2014.

[3] Nenni D. and McLellan P., *Fabless: The Transformation of the Semiconductor Industry*, CreateSpace Independent Publishing Platform, Seattle, WA, 2014.

Author's Biography

APEK MULAY

Apek Mulay is a business and technology consultant with Mulay's Consultancy Services. He is author of two books *Mass Capitalism: A Blueprint for Economic Revival* and *Sustaining Moore's Law: Uncertainty Leading to a Certainty of IoT Revolution*. He pursued undergraduate studies in electronics engineering (EE) at the University of Mumbai in India, and has completed a master's degree in EE at Texas Tech University, Lubbock. He is sole author of a patent "Surface Imaging with Materials Identified by Colors," developed during his employment in Advanced CMOS technology development team at Texas Instruments Inc. Mr. Mulay has chaired technical sessions at the International Symposium for Testing and Failure Analysis (ISTFA) for consecutive years 2009 and 2010. He has also authored several articles in reputed publications, showing this wide expertise in macroeconomics, geo-politics, supply chains, business models, socio-economics, and microeconomics, relating these to the capital-intensive semiconductor industry.

USCIS approved his US permanent residency under the category of foreign nationals with extraordinary abilities in science and technologies, even though he did not pursue a Ph.D. in either engineering or economics. He contributes to several recognized publications, such as *EBN, Truthout.org, electronics.ca publications, EDFA* magazine, *Military & Aerospace Electronics Magazine* (MA&E), SEMI.org, LinkedIn, etc. His blog www.apekmulay.com has reached close to 2 million hits since he started blogging in May 2013.

Printed in the United States
by Baker & Taylor Publisher Services